Lecture Notes in Statistics

Edited by D. Brillinger, S. Fienberg, J. Gani, J. Hartigan, and K. Krickeberg

30

Jan Grandell

Stochastic Models of Air Pollutant Concentration

Springer-Verlag
Berlin Heidelberg New York Tokyo

Author

Jan Grandell
Department of Mathematics, The Royal Institute of Technology
10044 Stockholm, Sweden

Mathematics Subject Classification (1980): 60G10, 60G55, 60J25, 60K05, 62M09, 86A10

ISBN 3-540-96197-6 Springer-Verlag Berlin Heidelberg New York Tokyo
ISBN 0-387-96197-6 Springer-Verlag New York Heidelberg Berlin Tokyo

Printed in Germany

Printing and binding: Beltz Offsetdruck, Hemsbach/Bergstr.
2147/3140-543210

PREFACE

About fifteen years ago Henning Rodhe and I disscussed the calculation of residence times, or lifetimes, of certain air pollutants for the first time. He was interested in pollutants which were mainly removed from the atmosphere by precipitation scavenging. His idea was to base the calculation on statistical models for the variation of the precipitation intensity and not only on the average precipitation intensity. In order to illustrate the importance of taking the variation into account we considered a simple model – here called the Markov model – for the precipitation intensity and computed the distribution of the residence time of an aerosol particle.

Our expression for the average residence time – here formula (13) – was rather much used by meteorologists. Certainly we were pleased, but while our ambition had been to provide an illustration, our work was merely understood as a proposal for a realistic model. Therefore we found it natural to search for more general models. The mathematical problems involved were the origin of my interest in this field.

A brief outline of the background, purpose and content of this paper is given in section 1.

It is a pleasure to thank Gunnar Englund, Georg Lindgren, Henning Rodhe and Michael Stein for their substantial help in the preparation of this paper and Iren Patricius for her assistance in typing.

Stockholm, December 1984 Jan Grandell

CONTENTS

1 INTRODUCTION

The purpose of this paper is to summarize and develop mathematical models for describing the variability of the concentration of pollutants in the atmosphere. More precisely we shall consider the time variability of the concentration of some material in the atmosphere. Let $c(t)$ (kg/m^3) be the concentration of the material at time t. The value of $c(t)$ fluctuates over time. We shall consider models where this fluctuation is due to variation in the source, i.e. in the rate of emission into the atmosphere, and the sink, i.e. the rate of removal. This means that the material enters and leaves the atmosphere with varying rates. Mathematically the source and the sink are described by stationary stochastic processes, and, within the simple physical model to be used, the concentration will also be a stationary stochastic process. We wish to point out that the situation is very different if spatial variability is considered. Then the concentration is measured at the same time at different locations. In such a case one may let the source and the sink be deterministic and the fluctuation in the concentration from location to location may be due only to geographical variations. The mathematically satisfying approach would be to consider $c(t,x)$, i.e., the concentration at time t at location x, which should in principle permit a simultaneous study of temporal and spatial variations. Such a study should, among other things, require a meteorological model for the relation between the air movement and the sink. To our knowledge there exist no such models which are realistic and simple enough to be suitable for our kind of analysis.

Let us consider the concentration of some kind of aerosol particles. An interesting quantity is the residence time; i.e., the time spent in the atmosphere by an individual particle before its removal. Rodhe and Grandell (1972) computed the residence time when the sink was described by a specific model. Independently Gibbs and Slinn (1973) considered a model for the concentration where both the sink and the source were described by stochastic processes. They described the processes in general terms and derived approximate results about the variability of the concentration. Mathematically, it is much simpler to derive results about the residence time than about the variability of the concentration. In practice, it is easy to measure the concentration but difficult or impossible to measure the residence time. Therefore it is interesting to relate properties of the concentration and the residence time. If the sink and the source are independent then the mean concentration equals the mean source rate times the mean residence time. Especially for natural sources, i.e., for sources not due to human activities, the mean source rate is difficult to estimate, and in such cases this simple relation is of little or no use. Intuitively, however, the concentration of particles with long residence time ought to vary less than for particles with short residence time. Thus it is interesting to relate residence times and variability.

After the appearence of the above mentioned works many papers containing different extensions have been published. Hopefully we are fairly complete with regard to the mathematical developements within the physical model proposed by Gibbs and Slinn (1973). We will, however, point out that we do not at all discuss extensions of the physical model like allowing for sink by diffusion, taking chemical transformations of the material into account and so on.

As mentioned this paper is about mathematical methods. The most important consequence is that many - important and difficult - "practical" questions are disregarded. Our hope is, however, that the rather generally formulated models and results might be useful in more specific situations. We have therefore tried to make the paper also readable for meteorologists with a good knowledge of probability theory. Therefore, proofs, technical derivations and formulations and comments of purely mathematical interest are collected in the appendices or in the mathematical remarks at the end of the relevant section.

In section 2 we state some notation and some basic facts of stochastic processes. A probabilist may of course omit it completely and maybe go back to it if some notation is unclear. Particularly we consider certain facts regarding point processes which sometimes are misunderstood in applications. In section 3 we formulate the model for $c(t)$ and discuss its interpretation. In section 4 we consider the mean concentration and its relation to residence times. In section 5 we consider the variance of $c(t)$ when the sink and the source are independent. Section 6 is devoted to the approximation proposed by Gibbs and Slinn (1973). In section 7 we consider precipitation scavenging, i.e. when the sink is related to the precipitation intensity. This section differs somewhat from the rest of this paper, in that more practical questions are discussed. In section 8 we consider some "stochastic process properties" of $c(t)$ and approximations of the distribution of the concentration. Among the five appendices we shall here only mention appendix A4 where models with dependent sink and source are considered.

2 SOME BASIC PROBABILITY

Let X be a random variable. Its _distribution_ _function_ F_X is defined by $F_X(x) = \Pr\{X \leq x\}$. Its _mean value_ is denoted by $E(X)$ or μ_X and its variance by $\mathrm{Var}(X)$ or σ_X^2. All basic random variables, considered in this paper, will be _non-negative_, i.e. $\Pr\{X < 0\} = 0$. Assume now that X is non-negative. Its _survivor_ _function_ G_X is defined by

$$G_X(x) = \Pr\{X > x\} = 1 - F_X(x)$$

and we have the relation

$$E(X) = \int_0^\infty x\,dF_X(x) = \int_0^\infty G_X(x)\,dx.$$

The _coefficient_ _of_ _variation_ V_X is defined by

$$V_X = \sigma_X / \mu_X.$$

Often $V_X^2 = \mathrm{Var}(X)/E(X)^2$ is called the _relative_ _variance_, and this will be our main measure of variability. For any fixed number α we have $V_{\alpha X} = V_X$.

Let $X(t)$ be a non-negative _stationary_ stochastic process. This means, for any n and any $t_1, \ldots, t_n$, that the distribution of $(X(t_1+h), X(t_2+h), \ldots, X(t_n+h))$ does not depend on h. Because of of stationarity $E(X(t))$ and $\mathrm{Var}(X(t))$ do not depend on t. Since there will be no risk for missunderstanding we use the notation $\sigma_X^2 = \sigma_{X(t)}^2$ and $V_X = V_{X(t)}$. The _covariance_ _function_ r_X, defined by

$$r_X(\tau) = \mathrm{Cov}(X(t), X(t+\tau))$$

does, again by stationarity, not depend on t. In order to avoid technical complications we always assume that

$$\int_{-\infty}^\infty |r_X(\tau)|\,d\tau < \infty.$$

Then r_X has the representation

$$r_X(\tau) = \int_{-\infty}^{\infty} e^{i\tau\omega} f_X(\omega)\,d\omega$$

and conversely

$$f_X(\omega) = \frac{1}{2\pi} \int_{-\infty}^{\infty} e^{-i\tau\omega} r_X(\tau)\,d\tau$$

where $f_X(\omega)$ is called the <u>spectral density</u>. Further $f_X(\omega)$ is continuous and bounded.

Whenever we consider two stationary prosesses $X_1(t)$ and $X_2(t)$ it is understood that they are <u>simultaneously stationary</u>. Formally this means, for any n and any $t_1, \ldots, t_n$, that the distribution of $(X_1(t_1+h), X_2(t_1+h), \ldots, X_1(t_n+h), X_2(t_n+h))$ does not depend on h. Two independent stationary processes are always simultaneously stationary.

Under mild conditions, see for example Grandell (1982, p. 242), we can define a stochastic process Y(t) with stationary increments by

$$Y(t) - Y(s) = \int_s^t X(\tau)\,d\tau \quad \text{and} \quad Y(0) = 0.$$

In this case we talk about an <u>intensity model</u>, and it may be noted that $E(Y(1)) = E(X(t))$.

In our applications X(t) will be the model for the scavenging intensity, the precipitation intensity or the source strength. If X(t) represents the precipitation intensity, then $Y(t) - Y(s)$ is the total amount of precipitation in the time interval (s,t). In this case it may be reasonable to disregard the length of the precipitation event. We are then led to a class of processes with <u>stationary increments</u> which are not intensity models. Two processes with stationary increments are understood to have simultaneously stationary increments.

Let N(t) be a <u>stationary point process</u>, i.e., a process with stationary increments such that N(t) is constant everywhere except at isolated epochs where it increases exactly one unit. We shall consider stationary point processes somewhat more detailly than really necessary for this paper, since they are often erroneously used in applications.

Consider an arbitrary fixed time point, which we, for notational reasons, choose as the origin. In Figure 1 we consider a stationary point process. The crosses on the time axis indicate the times of the events. In the precipitation case the crosses thus illustrate the times of the showers.

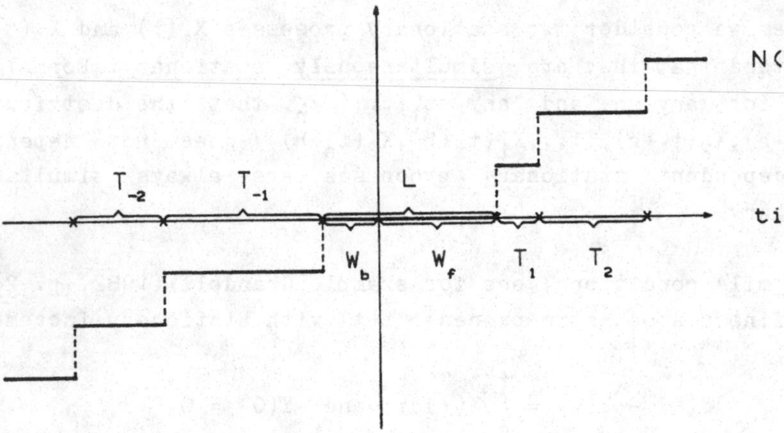

FIGURE 1: Illustration of some notation.

Now we restrict ourselves to the case where $L, T_1, T_{-1}, T_2, T_{-2}, \ldots$ are independent and where all T:s have a common distribution function F_T. In this case N(t) is a <u>stationary</u> <u>renewal</u> <u>process</u>. It is well-known that

$$F_{W_f}(x) = F_{W_b}(x) = \frac{1}{\mu_T} \int_0^x G_T(y)\,dy \tag{1}$$

and thus

$$E(W_f) = (\sigma_T^2 + \mu_T^2)/(2\mu_T) \tag{2}$$

where $\mu_T = E(T_k)$ and $\sigma_T^2 = \mathrm{Var}(T_k)$. One error, usual in the applications of point processes, is that W_f and T_k are not properly separated. Usually this means that W_f, which is the time from an arbitrary fixed time point until the next event, and any T_k, which is the time from an event until the next event, are treated as identically distributed. From (1) it follows that $F_{W_f} = F_T$ if and only if T_k is exponentially distributed; i.e., if

$$F_T(x) = 1 - e^{-x/\mu_T} \qquad \text{for} \quad x \geq 0.$$

In that case N(t) is a <u>Poisson process</u>. The Poisson process is the only point process with stationary <u>and independent</u> increments. It is not unusual in applications that the dependence between increments is disregarded, which thus implies that a Poisson process is implicitly assumed, although a greater generality is claimed.

Let Y(t) be a process with stationary increments. If it, for some stationary point process N(t), has the representation

$$Y(t) - Y(s) = \sum_{k=N(s)+1}^{N(t)} \tilde{X}_k$$

we talk about a <u>point process</u> model, or, when Y(t) - Y(s) is the total amount of precipitation in (s,t], a <u>short rain model</u>. If N(t) is a Poisson process and the $\tilde{X}$:s are independent of each other and of N(t) and if all $\tilde{X}$:s have a common distribution function we talk about a <u>Poisson model</u>. If N(t) is a stationary renewal process, and the $\tilde{X}$:s are as in the Poisson model, we talk about a <u>S. R. renewal model</u>. (The "S. R." stands for short rain. In Rodhe and Grandell (1981) the name renewal model was used for a certain intensity model.)

Let Y(t) be a process with stationary increments. Assume that there exists a constant Γ_Y, defined by

$$\Gamma_Y = \lim_{t\to\infty} \mathrm{Var}(Y(t))/t.$$

This means that we assume that Var(Y(t)) is asymptotically linear in t. In an intensity model we have

$$\Gamma_Y = \int_{-\infty}^{\infty} r_X(\tau)d\tau = 2\pi f_X(0).$$

In a point process model with independent $\tilde{X}$:s, like in the Poisson model, we have

$$E(Y(1)) = E(\tilde{X})E(N(1))$$

and

$$\Gamma_Y = Var(\tilde{X})E(N(1)) + E(\tilde{X})^2\Gamma_N.$$

For a stationary renewal process we have $E(N(1)) = 1/\mu_T$ and $\Gamma_N = \sigma_T^2/\mu_T^3$ and thus we have in the S. R. renewal model

$$E(Y(1)) = E(\tilde{X})/\mu_T$$

and

$$\Gamma_Y = (Var(\tilde{X})/\mu_T) + E(\tilde{X})^2\sigma_T^2/\mu_T^3 = (Var(\tilde{X}) + E(\tilde{X})^2V_T^2)/\mu_T.$$

For the Poisson process this reduces to

$$\Gamma_Y = (Var(\tilde{X}) + E(\tilde{X})^2)/\mu_T = E(\tilde{X}_k^2)/\mu_T$$

and in this case we have $Var(Y(t)) = t\Gamma_Y$.

One of the simplest intensity models is the <u>two-state</u> <u>Markov process</u>. In that case $X(t)$ changes between two states x_d and x_p. The Markov property implies that the lengths of the periods in the states are independent and exponentially distributed. Let τ_d and τ_p be the mean length of a period in state x_d and x_p respectively. Due to stationarity we have

$$p_d = Pr\{X(t) = x_d\} = \tau_d/(\tau_d + \tau_p)$$

$$p_p = Pr\{X(t) = x_p\} = \tau_p/(\tau_d + \tau_p).$$

Further we have

$$x_0 = E(X(t)) = x_d p_d + x_p p_p$$

$$r_X(\tau) = (x_d - x_p)^2 p_d p_p e^{-A_X|\tau|} = (x_0 - x_d)^2 \frac{p_d}{p_p} e^{-A_X|\tau|}$$

$$f_X(\omega) = \frac{(x_d - x_p)^2 p_d p_p A_X}{\pi(A_X^2 + \omega^2)} = \frac{(x_0 - x_d)^2}{\pi p_p \tau_p (A_X^2 + \omega^2)}$$

where

$$A_X = \frac{1}{\tau_0} + \frac{1}{\tau_1} .$$

3 THE GENERAL MODEL

As indicated in the introduction we shall consider time variability. More precisely we consider a stationary stochastic process $c(t)$ (kg/m^3) describing the concentration of some kind of aerosol particles or some trace gas. Assume that

$$c'(t) = Q(t) - c(t)\lambda(t) \tag{3}$$

where $Q(t)$ (kg/m^3h) and $\lambda(t)$ (h^{-1}) are stationary processes called the <u>source strength</u> and the <u>sink intensity</u>. This simple physical model was first proposed by Gibbs and Slinn (1973), and we shall essentially restrict ourselves to that model. For particles in the size range 0.1 - 1 μm and water-soluble gases we interpret the sink intensity as the precipitation scavenging intensity. Let the stationary process $R(t)$ (kg/m^2h) describe the <u>precipitation intensity</u>. In this case we shall sometimes assume that $\lambda(t) = aR(t)$ where a (m^2/kg) is a "parameter" depending on the type and the size of the particles or on the kind of the gas. In many applications the value of a ranges from 0.1 to 1.

The interpretation of $\lambda(t)$, $Q(t)$ and thus $c(t)$ must be in a Lagrangian sense. This means that $\lambda(t)$ and $Q(t)$ shall be models for the sink intensity and the source strength acting on a system moving with the winds. If we consider precipitation scavenging we can easily get precipitation measurements from fixed stations – i.e. Eulerian data – but it is difficult, or maybe impossible, to get Lagrangian type data. Since there are indications, see Hamrud et al. (1981), that Lagrangian and Eulerian rainfall data do not differ too much, and since in this paper we consider mathematical aspects, we disregard this problem. When we consider the source strength the situation is different. If the particles or the gas are produced by human activities it is natural to regard the variability as mainly induced by the movement of the system. To be concrete we may think of an area with rather isolated sources. The value of $Q(t)$ does then mainly depend on whether the "air parcel" is close to a source or not.

If we disregard the variation in the source strength, i.e., we put $Q(t) = Q_0$ where Q_0 is a fixed number, we might consider $c(t)$ in an Eulerian sense. There is, however, still an important complication, namely that the concentration does not only depend on the sink intensity and the source strength, but also on the transport of the particles or the gas. This means that we must consider $c(t)$ as the average concentration in an area so large that the amount of particles or gas blowing into it and away from it is small compared to the amount emitted and removed within it.

From now on we disregard all "practical" problems and interpret all quantities in the Lagrangian sense. There is still a problem of interpreting $c(t)$ in "the Lagrangian sense". We shall consider "the variability of $c(t)$" and formally we shall consider

$$V_{c(0)} = \sqrt{\mathrm{Var}(c(0))}/E(c(0)).$$

(The restriction to t=0 is motivated by the stationary assumptions, which imply that $V_{c(t)} = V_{c(0)}$ for all t.) What shall be meant by "$V_{c(0)}$ in the Lagrangian sense?" To this general question we have no answer, and therefore we propose the following interpretation.

Consider a fixed location. At that location, at time t=0, an "air parcel" is picked out. Let $c(t)$ be the concentration in that particular air parcel at time t and let $c_E(t)$, where "E" stands for Euler, be the concentration at that location at time t. Probably $c_E(t)$ is more interesting than $c(t)$. Unfortunately we can not handle $c_E(t)$, and consequently our proposed interpretation gives no information about the relation between $c_E(t)$ and $c(t)$ except that $c_E(0) = c(0)$. Our interpretation destroys the stationarity of $c(t)$, since $Q(0)$ is (almost) deterministic. One, at least theoretically possible, way to overcome this problem is to consider a stationary process $Q(t)$ and then condition upon $Q(t)$ for $t = 0$ or for t close to zero. We shall give an example in section 5 where this way is possible. Intuitively this means that $Q(t)$ can be regarded as approximately stationary except for t close to zero. Again, we shall disregard this complication and assume $c(t)$ to be stationary. It seems reasonable to make this simplification if the residence time of the particles is long compared to the "memory" of $Q(t)$, since then most particles are emitted a rather long time ago where

Q(t) is approximately stationary. Further the "fixed location" ought to be situated on a place which is "normal" compared to the surrounding area. In applications the choice of the process Q(t) must depend on the "fixed location" and its surrounding area.

Let us go back to formula (3). This formula only works when the source strength and the sink are described by intensity models. As indicated in section 2 it is sometimes natural to use point process models, and therefore we shall now give a general definition of the concentration process c(t).

In order to be concrete, we consider first precipitation scavenging. In the case of an intensity model we define the precipitation process h(t) (kg/m^2) by

$$h(t) = \int_0^t R(\tau)d\tau.$$

This implies that $h(0) = 0$ and that $h(t) - h(s) = \int_s^t R(\tau)\, d\tau$, i.e., $h(t) - h(s)$ is the total amount of precipitation in the interval (0,t]. Thus the precipitation process h(t) is a process with stationary increments, and it may very well be a point process model. If we, again, assume an intensity model it is sometimes reasonable to assume that $\lambda(t) = a\, R(t)$. In this case we define the the sink process $\Lambda(t)$ by

$$\Lambda(t) = \int_0^t \lambda(\tau)d\tau = a \int_0^t R(\tau)d\tau = a\, h(t).$$

Thus $\Lambda(t) = a\, h(t)$ is a natural definition also when h(t) does not have an intensity representation. We shall, from now on, talk about the sink process $\Lambda(t)$ also when no special reference is made to precipitation scavenging.

We now define the source process q(t) (kg/m^3) by

$$q(t) = \int_0^t Q(\tau)d\tau$$

in case of an intensity model. This implies that $q(0) = 0$ and that $q(t) - q(s)$ is the concentration emitted into the air parcel in the interval (s,t]. Consider now c(t). In the time interval (s-ds,s],

$s < t$, the concentration $dq(s)$, which equals $Q(s)ds$ in case of of an intensity model, is emitted into the air parcel. The fraction of $dq(s)$ which remains in the atmosphere at time t is determined by the sink process. As we have defined the sink process, this fraction equals

$$e^{-(\Lambda(t) - \Lambda(s))} \tag{4}$$

and thus

$$c(t) = \int_{-\infty}^{t} e^{-(\Lambda(t) - \Lambda(s))} dq(s) \tag{5}$$

and this is our general definition which is valid for any sink and source processes. When nothing else is said the sink and source processes are assumed to have simultaneously stationary increments which implies the stationarity of $c(t)$.

In the case of intensity models we have $dq(s) = Q(s)ds$ and $\Lambda'(t) = \lambda(t)$. Thus

$$c'(t) = e^{-(\Lambda(t) - \Lambda(t))}Q(t) - \int_{-\infty}^{t} \lambda(t)e^{-(\Lambda(t) - \Lambda(s))}Q(s)ds =$$

$$= Q(t) - c(t)\lambda(t)$$

which coincides with (3). Consider now (4) and put

$$\xi(t) = e^{-\Lambda(t)}. \tag{6}$$

Thus $\xi(t)$ may be interpreted as the concentration remaining at time t of a unit concentration emitted at time 0. In the case of an intensity model this interpretation is well-known, but we shall give a short indication of it. The sink intensity $\lambda(t)$ means that the fraction $\lambda(t)dt$ is removed from the atmosphere in the time interval $(t,t+dt]$. Thus $\xi(t+dt) = \xi(t)(1 - \lambda(t)dt)$ which implies that

$$\xi'(t) = - \lambda(t)\xi(t)$$

and thus $\xi(t)$ is given by (6) since $\xi(0) = 1$. Consider now a point process model. Then, see section 2,

$$\Lambda(t) = \sum_{k=1}^{N(t)} \tilde{\lambda}_k \quad \text{and thus} \quad \xi(t) = \prod_{k=1}^{N(t)} e^{-\tilde{\lambda}_k}$$

where the random variable $\tilde{\lambda}_k$ is the sink at "shower" number k. We always make the convention that $\sum_{k=1}^{0} \ldots = 0$ and $\prod_{k=1}^{0} \ldots = 1$.

Thus $e^{-\tilde{\lambda}_k}$ is the fraction of the concentration left by "shower" number k. If we were only interested in point process models, we could of course make a model for the fraction removed by "shower" number k. Call this fraction ε_k. Then $\xi(t) = \prod_{k=1}^{N(t)} (1 - \varepsilon_k)$ and this notation is used by Slinn (1982, p. 58), for example.

Consider a single particle - or gas molecule - which enters the atmosphere at time 0. Its <u>residence time</u> in the atmosphere is a random variable T. In the case of $\Lambda(t) = a\,h(t)$ we denote the residence time by T_a. If we think of that particle as one of many emitted at time 0 the probability that it remains in the atmosphere at time t, i.e. that $T > t$, is $\xi(t)$. If the particle enters the atmosphere independently of the sink process we thus get

$$G(t) = \Pr\{T > t\} = E(\xi(t)) = E(e^{-\Lambda(t)})$$

and

$$E(T) = \int_0^\infty G(t)dt$$

For a rather detailed treatment of residence times we refer to Rodhe and Grandell (1981). We shall return to residence times in section 4.

A major purpose of this paper is to relate $V_{c(0)}$ and $E(T)$. Recall that

$$V_{c(0)}^2 = \frac{Var(c(0))}{E(c(0))^2} .$$

Intuitively, $V_{c(0)}$ decreases with increasing $E(T)$. A knowledge about the relation between $V_{c(0)}$ and $E(T)$ is especially important for "long-lived" particles; i.e., when $E(T)$ is large. All results we know about are of the form

$$V_{c(0)} \propto E(T)^{\alpha}$$

where ∝ means "proportional to." Gibbs and Slinn (1973) found $\alpha = -1/2$ and their result is supported by later investigations for models of the kind treated in this paper. Using empirical data Junge (1974) found $\alpha = -1$ for the spatial variation. Hamrud (1983) also considered spatial variation and from numerical experiments he found α-values ranging from -0.92 to -0.72. Junge's relation has been used extensively as a tool to estimate the residence time. Hamrud (1983) found 44 papers referring to Junge (1974) up to 1981. In most of those papers the residence time was estimated from the variability of the concentration.

Mathematical remark

The differential equation (3) must be interpreted with some care. We always assume that the realizations of $\lambda(t)$ and $Q(t)$ are piecewise continuous and that the discontinuity points are at most finitely many on finite intervals. Further the realizations are assumed to be right-continuous. For each fixed value of t equation (3) holds with probability one. For given realizations of $\lambda(t)$ and $Q(t)$ equation (3) holds almost everywhere. In order to avoid trivial complications, we therefore let $c'(t)$ mean the right-hand derivative; i.e.,

$$c'(t) = \lim_{\Delta \downarrow 0} \frac{c(t+\Delta) - c(t)}{\Delta}.$$

In the general case we define $dq(t) = q(t) - q(t-dt)$, i.e. as left-hand differential. The reason is that if there is a jump in $q(t)$ at time t, then $dq(t)$ equals that jump. Formally we thus have, in an intensity model, $dq(t) = Q(t-)dt$ where $Q(t-)$ is the left-hand limit, i.e., $Q(t-) = \lim_{s \uparrow t} Q(s)$. We have chosen to give "local" definitions of derivatives and differentials since we believe that they are intuitively attractive. Strictly mathematically this discussion is irrelevant since derivatives may be changed at isolated points and differentials only occur in integrals.

Now we consider the general case. Put $\tilde{\lambda}(t) = \Lambda(t) - \Lambda(t-)$; i.e., $\tilde{\lambda}(t)$ is the jump of $\Lambda(t)$ at time t. Differentiation of (5) yields, see e.g., Elliott (1982, p.132),

$$dc(t) = e^{-\Lambda(t)}\, d\left(\int_{-\infty}^{t} e^{\Lambda(s)} dq(s)\right) + d(e^{-\Lambda(t)}) \int_{-\infty}^{t-} e^{\Lambda(s)} dq(s) =$$

$$= dq(t) - c(t-)(d\Lambda(t) - \tilde{\lambda}(t) + 1 - e^{-\tilde{\lambda}(t)})$$

and this is the general version of (3). For intensity models we have $c(t-) = c(t)$ and $\tilde{\lambda}(t) = 0$ and we are back to (3). For point process models we have $d\Lambda(t) = \tilde{\lambda}(t)$ and the general equation is reduced to

$$dc(t) = dq(t) - c(t-)(1 - e^{-\tilde{\lambda}(t)}).$$

Recall that $dq(t)$ is the concentration <u>emitted into the air parcel</u> while it might be more natural to consider models for the concentration <u>emitted from the earth</u>. The difference is that some fraction of the concentration emitted from the earth might not remain in the atmosphere long enough to reach the air parcel. Since we have no models for the time it takes for a particle to reach the air parcel we formally put this time equal to zero. This means that the two interpretations of $dq(t)$ coincide provided $\Lambda(t)$ and $q(t)$ have no simultaneous jumps. This is the case if $\Lambda(s)$ and $q(t)$ are independent.

Let us now allow for simultaneous jumps. Assume that both $\Lambda(t)$ and $q(t)$ have jumps $\tilde{\lambda} = \tilde{\lambda}(t)$ and $\tilde{Q} = \tilde{Q}(t) = Q(t) - Q(t-)$ at time t. Consider the point process model as an approximation of an intensity model. Let the effect of $\tilde{\lambda}$ and $\tilde{Q}$ be evenly spread out over the interval $(t, t + \Delta)$. This corresponds to $\lambda(s) = \tilde{\lambda}/\Delta$ and $Q(s) = \tilde{Q}/\Delta$ for $t \le s \le t + \Delta$. The part of $\tilde{Q}$ remaining in the air parcel at time $t + \Delta$ is then

$$\int_0^{\Delta} e^{-\tilde{\lambda}(\Delta - s)/\Delta}\, (\tilde{Q}/\Delta)\, ds = \tilde{Q}(1 - e^{-\tilde{\lambda}})/\tilde{\lambda}.$$

Thus it seems natural to replace $dq(s)$ by $dq(s)(1 - e^{-\tilde{\lambda}(s)})/\tilde{\lambda}(s)$ in (5) if simultaneous jumps are allowed where $(1 - e^{-\tilde{\lambda}(s)})/\tilde{\lambda}(s)$ is interpreted as one if $\tilde{\lambda}(s) = 0$. We return to this in appendix A4.

4. RESIDENCE TIMES AND MEAN CONCENTRATIONS

The residence time T of a single particle is the time spent in the atmosphere of that particle. From the definitions in section 3 it follows that

$$E(c(t)) = \int_{-\infty}^{t} E(e^{-(\Lambda(t)-\Lambda(s))}dq(s)).$$

Put $c_0 = E(c(t))$, $\lambda_0 = E(\Lambda(1))$ and $Q_0 = E(q(1))$ and recall from section 2 that this implies, in the case of intensity models, that $\lambda_0 = E(\lambda(t))$ and $Q_0 = E(Q(t))$. Assume now that $\Lambda(t)$ and $q(s)$ are independent. Then

$$c_0 = \int_{-\infty}^{t} G(t-s)Q_0 ds = Q_0 \int_{0}^{\infty} G(s)ds = Q_0 E(T) \tag{7}$$

which is quite natural. Note that (7) does not hold in general if dependence is allowed between $\Lambda(t)$ and $q(s)$. An illustration of this fact is given in appendix A4.

The simplest model is the <u>classical</u> <u>model</u> where the variation in the sink process is disregarded; i.e., $\Lambda(t) = \lambda_0 t$. In this case

$$G(t) = e^{-\lambda_0 t} \tag{8}$$

and

$$E(T) = 1/\lambda_0. \tag{9}$$

In the general case, see appendix A1, we have

$$G(t) \geq e^{-\lambda_0 t}, \quad t \geq 0,$$

which implies

$$E(T) \geq 1/\lambda_0$$

and if $\Lambda(t)$ and $q(s)$ are independent we thus have

$$c_0 \geq Q_0/\lambda_0. \tag{10}$$

This implies that the mean concentration is systematically underestimated if the random variation in the sink process is disregarded.

If $\Lambda(t)$ and $q(s)$ are dependent, (10) is not always true. It can, however, be shown, see appendix A1, that

$$c_0 \geq \exp\{E(\log(Q(0)))\}/\lambda_0 \qquad (11)$$

provided the source is described by an intensity model. In the same way as Q_0 corresponds to an arithmetic mean $\exp\{E(\log(Q(0)))\}$ corresponds to a geometric mean. It shall be observed that if $\Pr\{Q(0) = 0\} > 0$ then (11) reduces to the triviality $c_0 \geq 0$.

Assume that both the sink and source are described by intensity models and that $\lambda(t)$ only takes the values 0 and $\lambda_p > 0$. From (3) it then follows that

$$E(c'(t)) = Q_0 - \Pr\{\lambda(t) = \lambda_p\}\lambda_p\, E(c(t)|\lambda(t) = 0).$$

Under general assumption, see appendix A2, we thus have

$$E(c(0)|\lambda(0) = \lambda_p) = Q_0/\lambda_0. \qquad (12)$$

Thus, if $\lambda(t)$ and $Q(s)$ are independent, the underestimation of $E(c(0))$ in the classical model corresponds to the difference in the mean concentration at an arbitrary time and a time where $\lambda(t) = \lambda_p$.

We shall now consider some special models $\lambda(t)$. In the <u>Markov model</u>, considered by Rodhe and Grandell (1972), $\lambda(t)$ is assumed to be a two-state Markov process (see section 2) taking the values λ_d and λ_p. Let τ_d and τ_p be the mean length of a period in state λ_d and λ_p respectively. This model is considered in detail in appendix A4. We shall here only note that if $\lambda_d = 0$, then

$$E(T) = \frac{1}{\lambda_0} + p_d\tau_d \qquad (13)$$

where $p_d = \Pr\{\lambda(0) = 0\} = \tau_d/(\tau_d+\tau_p)$. Due to the properties of the exponential distribution (13) also follows from (12).

Rodhe and Grandell (1981) proposed a generalization where the length of a period in state λ_d has an arbitrary distribution with mean τ_d and variance σ_d^2. Then, again if $\lambda_d = 0$, it follows from Grandell and Rodhe (1978) that

$$E(T) = \frac{1}{\lambda_0} + \frac{p_d(\sigma_d^2 + \tau_d^2)}{2\tau_d} \tag{14}$$

Due to (2), (14) also follows from (12).

Now we consider point process models, that is when

$$\Lambda(t) = \sum_{k=1}^{N(t)} \tilde{\lambda}_k \qquad \text{for } t \geq 0$$

and

$$\Lambda(t) = -\sum_{k=N(t)+1}^{0} \tilde{\lambda}_k \qquad \text{for } t > 0,$$

and assume that the random variables $\tilde{\lambda}_k$:s are independent of each other and of $N(t)$ and that all $\tilde{\lambda}_k$:s have the same distribution.

Let $\phi(u)$ be the Laplace-transform corresponding to $\tilde{\lambda}_k$; i.e.,

$$\phi(u) = E(e^{-u\tilde{\lambda}_k}).$$

Thus $\phi(1)$ is the probability for a particle to "survive" a "shower." In the Poisson model we have

$$G(t) = \exp\{-t(1 - \phi(1))/\tau_d\} \tag{15}$$

and thus

$$E(T) = \tau_d/(1 - \phi(1)).$$

In the S.R. renewal model we can not express $G(t)$ exactly in an explicit way. Under general conditions on the distribution of the time $\tilde{T}_d$ between two "showers", see Grandell (1982, p. 247) for details, we have

$$\lim_{t\to\infty} e^{\kappa t}G(t) = \frac{((1-\phi(1))/(\kappa\phi(1)))^2}{\tau_d E(\tilde{T}_d \exp(\kappa\tilde{T}_d))} \tag{16}$$

where κ is the solution of

$$\phi(1)E(\exp\{x\tilde{T}_d\}) = 1.$$

Further we have

$$E(T) = \frac{\sigma_d^2 - \tau_d^2}{2\tau_d} + \frac{\tau_d}{1-\phi(1)}. \tag{17}$$

Now consider "long-lived" particles. Let $h(t)$ be a process with stationary increments and put $R_0 = Eh(1)$ and $\Gamma_h = \lim_{t\to\infty} t^{-1}\mathrm{Var}(h(t))$. Assume that $\Lambda(t) = ah(t)$ and denote the residence time by T_a. Under general assumptions, see appendix A3, for details, we have

$$G_a(t/a) \simeq e^{-R_0 t}\left(1 + \frac{at\Gamma_h}{2}\right) \tag{18}$$

where $G_a(t) = \Pr\{T_a > t\}$ and

$$E(T_a) \simeq \frac{1}{aR_0} + \frac{\Gamma_h}{2R_0^2} \tag{19}$$

for small values of a. As an alternative of (20) we have

$$G_a(t/a) \simeq \exp(-tR_0 + at\Gamma_h/2). \tag{20}$$

Assume now that $\Lambda(t)$ is "approximately normal" for large values of t, which holds for all specific models under consideration. Formally this means that

$$(\Lambda(t) - \lambda_0 t)/\sqrt{t\Gamma_\Lambda} \overset{d}{\simeq} W \tag{21}$$

where " $\overset{d}{\simeq}$ " means "approximately the same distribution as" and where W is a normally distributed random variable with $E(W) = 0$ and $\mathrm{Var}(W) = 1$. This implies that

$$\xi(t) = e^{-\Lambda(t)} \overset{d}{\simeq} \exp(-t\lambda_0 - W\sqrt{t\Gamma_\Lambda}) \tag{22}$$

and thus $\xi(t)$ is approximately log-normally distributed for large values of t. This interesting observation is due to Slinn (1982, p. 59). It is tempting, but mathematically <u>not</u> justified, to take the mean value of both sides in (22). If this anyhow is done, we are led to

$$G(t) \simeq \exp(-t\lambda_0 + t\Gamma_\Lambda/2). \tag{23}$$

This is, however, <u>not</u> a reasonable approximation. To realize that we consider the <u>S. R. Markov</u> model, i.e., a Poisson model where the $\tilde{\lambda}_k$:s are exponentially distributed with mean $\lambda_0\tau_d$. Then $\Gamma_\Lambda = 2\lambda^2\tau_d$ and

$$G(t) = \exp\{-t\lambda_0/(1 + \lambda_0\tau_d)\} \tag{24}$$

Thus (23) implies

$$\exp\{-t\lambda_0/(1 + \lambda_0\tau_d)\} \simeq \exp\{-t\lambda_0(1 - \lambda_0\tau_d)\}$$

and thus (23) may be reasonable only if

$$1/(1 + \lambda_0\tau_d) \simeq 1 - \lambda_0\tau_d$$

or, which is the same, if $\lambda_0\tau_d \simeq 0$. Thus we are back in the "long-lived" case. In our opinion an approximation like (23) is motivated only if it can be expressed like the approximation (16). Thus we do not regard this tempting approach as a possible way to derive useful approximations.

In section 8 we shall return to the case where $\Lambda(t)$ is assumed to be "approximately normal," and consider approximations of the concentration process itself.

<u>Mathematical remark</u>

Formulae (16) and (17) follow from the results in Grandell and Rodhe (1978), where an intensity model was considered. Since that case is much more complicated we shall give a direct proof of (17).

Using the notation in section 2 it follows that the time to the k:th "shower" is $W_f + T_1 + \ldots + T_{k-1}$ and thus its mean value is

$$(\sigma_d^2 + \tau_d^2)/2\tau_d + (k-1)\tau_d \ .$$

The probability that the particle is removed in the k´th "shower" is

$$\phi^{k-1}(1)(1 - \phi(1))$$

and thus

$$E(T) = (\sigma_d^2 + \tau_d^2)/(2\tau_d) + (1 - \phi(1)) \sum_{k=0}^{\infty} k\phi^k(1)\tau_d =$$

$$= (\sigma_d^2 + \tau_d^2)/(2\tau_d) + \tau_d\phi(1)/(1 - \phi(1)) =$$

$$= (\sigma_d^2 - \tau_d^2)/(2\tau_d) + \tau_d/(1 - \phi(1)).$$

5 THE VARIANCE OF THE CONCENTRATION

We shall, in this section consider, $Var(c(0))$ in the case where $\Lambda(t)$ and $q(s)$ are independent. To our knowledge, this general situation was first studied by Gibbs and Slinn (1973) who proposed an approximation for $Var(c(0))$. We shall consider their approach in section 6. Baker et al. (1979) and Grandell (1982) considered the case with random sink and deterministic source. Baker et al. (1979) also considered the "inverse" case with deterministic sink and random source. Lozowski (1983) considered the case where both the sink and source are independent and random.

It follows from (5) that

$$c(0) = \int_{-\infty}^{0} e^{\Lambda(x)} dq(x). \tag{25}$$

Let $c_{Q_0}(0)$ be the concentration when $q(s)$ is replaced by its mean value $Q_0 s$, and thus $c_{Q_0}(0)$ is computed as if the source were deterministic. Recall from (7) that $c_0 = E(c(0)) = E(c_{Q_0}(0)) = Q_0 E(T)$ where T is the residence time. From (5), it follows that

$$Var(c(0)) = E(c^2(0)) - c_0^2 = \int_{-\infty}^{0}\int_{-\infty}^{0} E(e^{\Lambda(x)+\Lambda(y)})E(dq(x)dq(y)) - c_0^2 =$$

$$= Var(c_{Q_0}(0)) + \int_{-\infty}^{0}\int_{-\infty}^{0} E(e^{\Lambda(x)+\Lambda(y)})\ Cov(dq(x),dq(y))$$

where $Cov(dq(x),dq(y)) = E(dq(x),dq(y)) - Q_0^2 dxdy$.

Now we assume that

$$Cov(dq(x),dq(y)) = r_q dxd\delta(y-x) + r_Q(y-x)dxdy \tag{26}$$

where $\delta(s) = \begin{cases} 0 & \text{if} \quad s < 0 \\ 1 & \text{if} \quad s > 0 \end{cases}$ and $r_q dx = E((dq(x))^2)$.

Thus we have

$$\mathrm{Var}(c(0)) = \mathrm{Var}(c_{Q_0}(0)) + r_q \int_{-\infty}^{0} E(e^{2\Lambda(x)})dx +$$

$$+\int_{-\infty}^{0}\int_{-\infty}^{0} E(e^{\Lambda(x)+\Lambda(y)})r_Q(y-x)dxdy =$$

$$= \mathrm{Var}(c_{Q_0}(0)) + r_q E(T^{(2)}) + 2\int_{-\infty}^{0}\int_{-\infty}^{0} E(e^{2\Lambda(x)}e^{\Lambda(x+y)-\Lambda(x)})r_Q(y)dydx \quad (27)$$

where $T^{(2)}$ is the residence time when $\Lambda(t)$ is replaced by $2\Lambda(t)$.

Let us now consider (26) in some detail. If the source is described by an intensity model then $r_q = 0$ and $r_Q(\tau)$ is the ordinary covariance function for $Q(t)$. Let now the source be described by a point process model where

$$q(t) = \sum_{k=1}^{N_q(t)} \tilde{Q}_k.$$

We shall always assume that the $\tilde{Q}_k$:s have a common distribution and are independent of each other and of $N_q(t)$. We call the times of of increase of $N_q(t)$ the <u>source times</u>. Let $\tilde{T}_q$ be the time between two source times. Let τ_q be the mean and $F_q(t)$ the distribution function of $\tilde{T}_q$. Then

$$r_q = E(\tilde{Q}_k^2)/\tau_q \quad (28)$$

and, for $\tau \neq 0$,

$$r_Q(\tau)d\tau = Q_0^2\tau_q\{E(dN_q(\tau)\,|\,dN_q(0) = 1) - \frac{d\tau}{\tau_q}\} \quad (29)$$

where $E(dN_q(\tau)\,|\,dN_q(0) = 1)$ is the mean number of source times in the interval $(\tau,\tau+d\tau)$ given that a source time occur at the origin. We shall return to (26), (28) and (29) in a mathematical remark The function $r_Q(\tau)$ is symmetric but it need not to be a covariance function.

If $q(t)$ is described by a Poisson model it has independent increments and then $r_Q(\tau) \equiv 0$.

Assume now that $N_q(t)$ is a stationary renewal process where F_q has density f_q and let f_q^{n*} be the density of the sum of n independent copies of $\tilde{T}_q$. Thus, f_q^{n*} is the density of the time from a source time until the n´th subsequent source time. Then, for $\tau > 0$,

$$r_Q(\tau) = Q_0^2 \tau_q \{ \sum_{n=1}^{\infty} f_q^{n*}(\tau) - \frac{1}{\tau_q} \}$$

which generally is difficult to compute. Let $\hat{f}_q(u)$ be the Laplace-transform of $f_q(\tau)$, i.e.,

$$\hat{f}_q(u) = E(e^{-u\tilde{T}_q}) = \int_0^{\infty} e^{-u\tau} f_q(\tau)d\tau,$$

and put

$$\hat{r}_Q(u) = \int_0^{\infty} e^{-u\tau} r_Q(\tau)d\tau.$$

Since $f_q^{n*}(u) = (\hat{f}_q(u))^n$ we have

$$\hat{r}_Q(u) = Q_0^2\tau_q \{ \frac{\hat{f}_q(u)}{1 - \hat{f}_q(u)} + \frac{1}{u\tau_q} \} \tag{30}$$

which often is simple to compute.

Now we consider $Var(c(0))$. First – compare (27) – we treat $Var(c_{Q_0}(0))$ for some specific models. In the classical model, i.e., $\Lambda(t) \equiv \lambda_0 t$, we have

$$Var(c_{Q_0}(0)) = 0$$

since no randomness is involved. In the Markov model, where $\lambda(t)$ is a two state Markov process, Baker et al. (1979) computed $Var(c_{Q_0}(0))$. The result is somewhat complicated and follows from (76) in appendix A4 if Q_d is put equal to Q_0. Their result simplifies to

$$\mathrm{Var}(c_{Q_0}(0)) = Q_0^2\tau_d p_d(\tau_d + \lambda_0^{-1} + p_p\tau_d)$$

if $\lambda_d = 0$. If $\lambda_d = 0$ but if the length of a period in state λ_d has an arbitrary distribution then $\mathrm{Var}(c_{Q_0}(0))$ follows from (83) in appendix A4 again by putting $Q_d = Q_0$. Stein (1984) has generalized the Markov model to a model where the sink and source are dependent and (83) is a very special case of his general model.

Now we consider point process models. Baker et al. (1979, p. 44) considered the Poisson model and showed that

$$\mathrm{Var}(c_{Q_0}(0)) = \frac{Q_0^2\tau_d^2}{1-\phi(1)}\left(\frac{2}{1-\phi(2)} - \frac{1}{1-\phi(1)}\right). \tag{31}$$

In the S. R. renewal model (31), see Grandell (1982, p.248), generalizes to

$$\mathrm{Var}(c_{Q_0}(0)) = \frac{Q_0^2\ \tau_d^2}{1-\phi(1)}\left(\frac{2}{1-\phi(2)} - \frac{1}{1-\phi(1)}\right) +$$

$$+ \frac{Q_0^2}{3\tau_d}\ (E(\tilde{T}_d^3) - 6\tau_d^3) - Q_0^2(\sigma_d^2 - \tau_d^2)\,\frac{1-2\phi(2)}{1-\phi(2)} - Q_0^2\left(\frac{\sigma_d^2-\tau_d^2}{2\tau_d}\right)^2. \tag{32}$$

Now we consider "long-lived" particles where $\Lambda(t) = ah(t)$ and $Q_0 = a\gamma_0$ for some constant $\gamma_0 > 0$. Under general assumptions, see appendix A3 for details we have

$$\mathrm{Var}(c_{Q_0}(0)) \approx \frac{a\gamma_0^2\Gamma_h}{2R_0^3} \tag{33}$$

for small values of a.

The next term in (27) is $r_q E(T^{(2)})$. This term is trivial in the sense that $E(T^{(2)})$ was discussed in section 4. Lozowsky (1983) considered the case where both the sink and the source were Poisson models and in that case we thus have

$$\mathrm{Var}(c(0)) = \frac{Q_0^2\ \tau_d^2}{1-\phi(1)}\left(\frac{2}{1-\phi(2)} - \frac{1}{1-\phi(1)}\right) + \frac{E(\tilde{Q}_k^2)}{\tau_q}\cdot\frac{\tau_d}{(1-\phi(2))}. \tag{34}$$

Consider now the last term

$$2\int_{-\infty}^{0}\int_{-\infty}^{0} E(e^{2\Lambda(x)}e^{\Lambda(x+y)-\Lambda(x)})r_Q(y)\,dy\,dx$$

in (27) and denote it by B. Let us consider the case where

$$E(e^{2\Lambda(x)}e^{\Lambda(x+y)-\Lambda(x)}) = \sum_{k=1}^{n} \alpha_k e^{\kappa_k^{(2)}x+\kappa_k y} \tag{35}$$

for $x,y < 0$. This case covers the classical model, the Poisson model and the Markov model. If $\Lambda(t) \equiv \lambda_0 t$ it is obvious that $n = 1$, $\alpha_1 = 1$, $\kappa_1 = \lambda_0$ and $\kappa_1^{(2)} = 2\lambda_0$. In the Poisson model it follows from (15), since $\Lambda(t)$ has independent increments, that $n = 1$, $\alpha_1 = 1$, $\kappa_1 = (1-\phi(1))/\tau_d$ and $\kappa_1^{(2)} = (1-\phi(2))/\tau_d$. In the Markov model it follows – from Rodhe and Grandell (1972) and the fact that $\Lambda(x)$ and $\Lambda(x+y) - \Lambda(x)$ are conditionally independent given $\lambda(x)$ – that (35) holds with $n = 4$.

Thus we have

$$B = 2\sum_{k=1}^{n} \alpha_k \int_{-\infty}^{0} e^{\kappa_k^{(2)}x}\,dx \int_{-\infty}^{0} e^{\kappa_k y} r_Q(y)\,dy = 2\sum_{k=1}^{n} \frac{\alpha_k}{\kappa_k^{(2)}}\hat{r}_Q(\kappa_k). \tag{36}$$

Assume now that $r_Q(\tau) = \sigma_Q^2 e^{-A_Q|\tau|}$. In the Poisson model (36) then reduces to

$$B = \frac{2\sigma_Q^2\tau_d^2}{(1-\phi(2))(1-\phi(1)+A_Q\tau_d)}. \tag{37}$$

In the S.R. renewal model, which unfortunately is not covered by (35), (37) generalizes to

$$B = \frac{2\sigma_Q^2}{A_Q}\left\{\frac{\sigma_d^2-\tau_d^2}{2\tau_d} + \frac{\tau_d}{1-\phi(2)} + \frac{1-\phi(1)}{A_Q(1-\phi(1)\hat{f}_d(A_Q))}\left(\frac{1-\hat{f}_d(A_Q)}{A_Q\tau_d} - \frac{1-\phi(2)\hat{f}_d(A_Q)}{1-\phi(2)}\right)\right\} \tag{38}$$

where $\hat{f}_d(A_Q) = E(\exp(-A_Q\tilde{T}_d))$. The proof of (38) is given in appendix A5.

Consider again "long-lived" particles where $\Lambda(t) = ah(t)$ and $q(t) = ag(t)$ for some non-decreasing process $g(t)$ with stationary increments. Put $\gamma_0 = E(g(1))$ and $\Gamma_g = \lim_{t\to\infty} \mathrm{Var}(g(t))/t$. Under general assumptions, see appendix A3, (22) generalizes to

$$\mathrm{Var}(c(0)) \approx \frac{a}{2R_0} \left(\frac{\gamma_0^2 \Gamma_h}{R_0^2} + \Gamma_g \right) \tag{39}$$

for small values of a.

Up to now we have always considered stationary sink and source. We shall now consider a simple example of a non-stationary source related to the fact – mentioned in section 3 – that the source sometimes is (almost) deterministic, close to $t = 0$. Let us, for simplicity assume the sink is deterministic, i.e., $\lambda(t) = \lambda_0$, and that the source is described by a – for the moment – stationary two-state Markov process $Q(t)$ taking the values 0 and 1. As usual we denote the corresponding concentration process by $c(t)$. The restriction to 0 and 1 is not too serious since if $\tilde{c}(t)$ is the concentration process when 0 and 1 is replaced by Q_d and Q_p, $Q_d < Q_p$, then $\tilde{c}(t) = Q_d/\lambda_0 + (Q_p - Q_d)c(t)$ and thus the restriction to 0 and 1 may merely be regarded as a way to make the notation simple.

Let τ_0 and τ_1 be the mean length of a period in state 0 and 1 respectively, and put $p_0 = \Pr\{Q(t) = 0\} = \tau_0/(\tau_0 + \tau_1)$ and $p_1 = 1 - p_0$. The transition probabilities are given by

$$p_{0,1}(x) = \Pr\{Q(x)=1 \mid Q(0)=1\} = p_1 - p_1 e^{A_Q x} \tag{40:a}$$

and

$$p_{1,1}(x) = \Pr\{Q(x)=1 \mid Q(0)=1\} = p_1 + p_0 e^{A_Q x} \tag{40:b}$$

for $x < 0$ where $A_Q = \dfrac{1}{\tau_0} + \dfrac{1}{\tau_1}$.

Now we leave the stationary case by assuming that $Q(0)$ is fixed or more generally that

$$\tilde{p}_0 = \Pr\{Q(0) = 0\}$$

but the (homogeneous) transition mechanism is kept. We shall only be interested in the cases $\tilde{p}_0 = 1$, $\tilde{p}_0 = p_0$ and $\tilde{p}_0 = 0$ which correspond to $Q(0) = 0$, the stationary case and $Q(0) = 1$ respectively. Put

$$d = \tilde{p}_0 - p_0$$

and

$$\tilde{p}_1(x) = \Pr\{Q(x)=1\} = \tilde{p}_0 p_{0,1}(x) + \tilde{p}_1 p_{1,1}(x) = p_1 - de^{A_Q x}$$

for $x < 0$. Thus, for $x,y < 0$, we get

$$Q_0(x) = E(Q(x)) = \tilde{p}_1(x)$$

and

$$r_Q(x,x+y) = \mathrm{Cov}(Q(x),\ Q(x+y)) = \tilde{p}_1(x)p_{1,1}(y) - \tilde{p}_1(x)\tilde{p}_1(x+y) =$$

$$= e^{A_Q y}(p_0 p_1 - d(p_0 - p_1)e^{A_Q x} - d^2 e^{2A_Q x}).$$

Thus

$$E(c(0) = \int_{-\infty}^{0} e^{\lambda_0 x}\, Q_0(x)dx$$

and, compare (27),

$$\mathrm{Var}(c(0)) = 2\int_{-\infty}^{0}\int_{-\infty}^{0} e^{2\lambda_0 x+\lambda_0 y}\, r_Q(x,x+y)dxdy.$$

Since the sink is deterministic we have $E(T) = 1/\lambda_0$. It follows after some integration that

$$E(c(0)) = K(\tau)\left(\frac{p_1\tau}{p_0} + 1 - \tilde{p}_0\right)$$

$$\mathrm{Var}(c(0)) = K^2(\tau)\left\{p_1(p_0+\tau) - \frac{2d(p_0+p_1)(p_0+\tau)}{2p_0+\tau} - d^2\right\}$$

where $\tau = E(T)/\tau_1$ and $K(\tau) = \tau p_0/(p_0 + \tau)$.

In Figures 2 and 3 we have put $\tau_1 = 1$ and consider $E(T)V^2_{c(0)}$ for $p_0 = 0.9$ and 0.7.

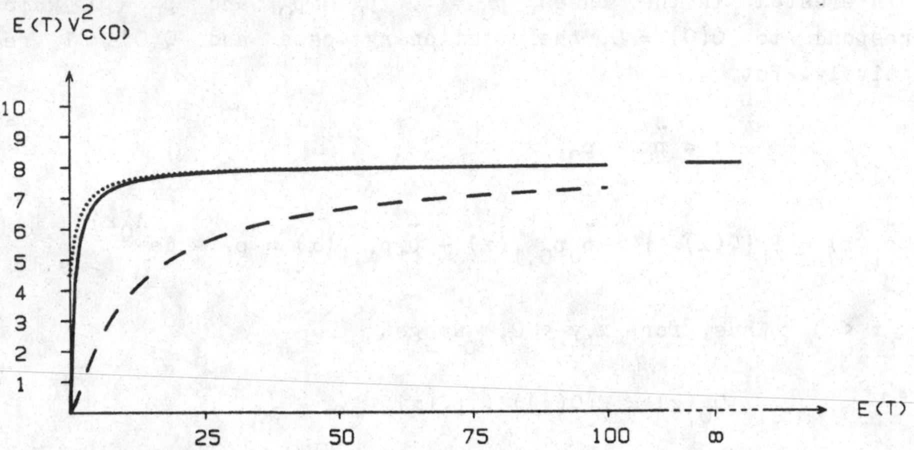

FIGURE 2: Illustration of the relative variance for varying values of Q(0) when $p_0 = 0.9$. From above the curves represent Q(0) = 0, the stationary case and Q(0) = 1 respectively.

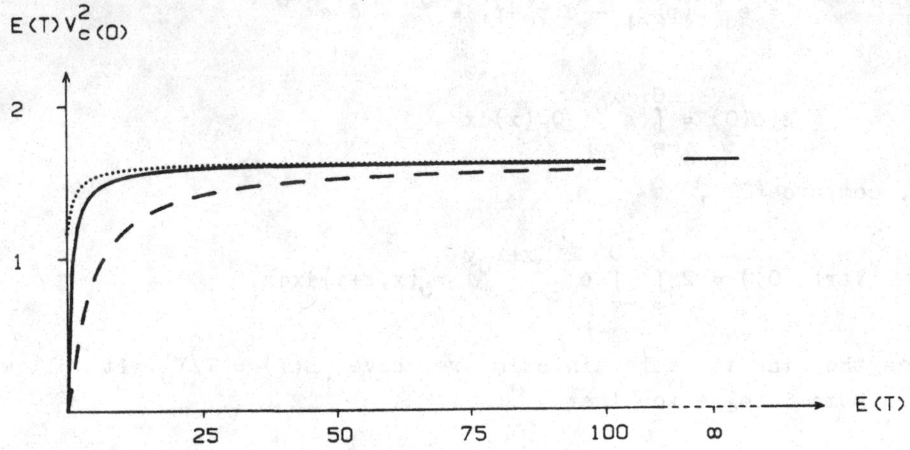

FIGURE 3: As Figure 2 but with $p_0 = 0.7$.

The figures indicate that it seems reasonable to assume Q(t) to be be stationary if E(T) is not too small and if Q(0) = 0. Thus this illustration supports the discussion in section 3, but it must be kept in mind that the model considered here is very special and probably not too realistic.

Mathematical remark

Let $q(t)$ be any non-decreasing process with stationary increments such that $E(q^2(t)) < \infty$. In order to formulate the general version of (26) we use the notation

$$q\{s+d\tau\} = q(s+\tau) - q(s+\tau-d\tau) \tag{41}$$

Mathematically $q\{\cdot\}$ is the random measure corresponding to the stochastic process $q(\cdot)$ and $d\tau$ is interpreted as the set $(\tau - d\tau, \tau]$. Note that $q\{d\tau\} = dq(\tau)$.

There always exists a signed measure R_Q with $dR_Q(0) = 0$, i.e., R_Q is the difference between two measures R_Q^+ and R_Q^-, such that

$$\mathrm{Cov}(q\{ds\}, q\{s+d\tau\}) = r_q\delta\{d\tau\}ds + R_Q\{d\tau\}ds. \tag{42}$$

Thus in (26) R_Q is assumed to be absolutely continuous.

From the definition of the covariance it follows that

$$\mathrm{Cov}(q\{ds\}, q\{s+d\tau\}) = E(q\{ds\}q\{s+d\tau\}) - Q_0^2 ds d\tau. \tag{43}$$

Consider now the case when $q(t)$ is a point process model. Since $N_q\{ds\} = 0$ or 1 it follows that $N_q^2\{ds\} = N_q\{ds\}$. Thus we have, for $\tau = 0$,

$$E(q\{ds\}q\{s+d\tau\}) = E(q^2\{ds\}) = E(\tilde{Q}_k^2)E(N_q^2\{ds\}) =$$

$$= E(\tilde{Q}_k^2)\, E(N_q\{ds\}) = E(\tilde{Q}_k^2)\, E(N_q(1))ds \tag{44}$$

and for $\tau \neq 0$

$$E(q\{ds\}q\{s+d\tau\}) = E(\tilde{Q}_k)^2\, E(N_q\{ds\}N_q\{s+d\tau\}). \tag{45}$$

Intuitively

$$E(N_q\{ds\}N_q\{s+d\tau\}) = E(N_q\{ds\})E(N_q\{s+d\tau\} \mid N_q\{ds\} = 1) =$$

$$= E(N_q(1))ds\, E(N_q\{d\tau\} \mid dN_q(0) = 1) \tag{46}$$

where the last equality is due to stationarity. In order to make the notion of "conditioned upon $dN_q(0) = 1$" precise we need the concept of Palm probabilities, and here we shall only indicate its definition. Let B be an event related to N_q; i.e., a set in the σ-algebra generated by N_q. Let B_x be the translation of B which means that the translated process $N_q(x+t) - N_q(x)$ belongs to B if and only if $N_q(t)$ belongs to B_x. Now define the "B-thinned" process N_q^B by

$$N_q^B(t) = \int_0^t 1_{B_x}(N_q) N_q\{dx\}$$

where

$$1_{B_x}(N_q) = \begin{cases} 1 & \text{if } N_q \text{ belongs to } B_x \\ 0 & \text{otherwise.} \end{cases}$$

The Palm probability $\Pr\{B \mid dN_q(0) = 1\}$ is defined by

$$\Pr\{B \mid dN_q(0) = 1\} = E(N_q^B(1)) \,/\, E(N_q(1)).$$

Let $\tilde{T}_q$ be a random variable with survivor function $G_q(t)$ given by

$$G_q(t) = \Pr\{N_q(t) = 0 \mid dN(0) = 1\}.$$

This is exactly what we mean when we in the text talk about "the time between two source times" or a "typical" period between source times.

Assume now that $\lim_{t\to\infty} \Pr\{N_q(t) = 0\} = 0$. For our applications this assumption seems quite harmless. It can be shown, cf. Matthes et al. (1978, p. 343) that $E(N_q(1)) = 1/\tau_q$ where as before $\tau_q = E(\tilde{T}_q)$, and thus (28) follows from (44). Further (43), (45) and (46) yield

$$R_Q\{d\tau\} = Q_0^2 \tau_q \; E(N_q\{d\tau\} \mid dN(0) = 1) - Q_0^2 d\tau$$

and thus (29) holds.

In section 7 we use the fact that (1) holds for all stationary point processes. Let $\tilde{W}_f$ be the time from zero until the next source time. Then - at least heuristically - we have, since $\Pr\{N_q\{ds\} = 1\} = E(N_q\{ds\}) = ds/\tau_q$,

$$\Pr\{\tilde{W}_f > t\} = \Pr\{N_q(t) = 0\} =$$

$$= \int_{-\infty}^{0} \Pr\{N_q(t) - N_q(s) = 0 | N_q\{ds\} = 1\} \Pr\{N_q\{ds\} = 1\} =$$

$$= \frac{1}{\tau_q} \int_{-\infty}^{0} G_q(t - s)ds = \frac{1}{\tau_q} \int_{t}^{\infty} G_q(s)ds.$$

A reader interested in these concepts is recommended to consult Daley and Vere-Jones (1972).

6 THE GIBBS AND SLINN APPROXIMATION

Assume that both the sink and the source are described by intensity models. Gibbs and Slinn (1973) derived an approximation for $V_{c(0)}$. The aim of this section is to consider their approximation in some detail. Note that $\lambda(t)$ and $Q(t)$ are now allowed to be dependent. Recall that $c(t)$, $\lambda(t)$ and $Q(t)$ are related by $c'(t)=Q(t)-\lambda(t)c(t)$. From this relation we get

$$(c(t)-c_0)' + \lambda_0(c(t)-c_0) =$$

$$= (Q(t) - Q_0) - c_0(\lambda(t)-\lambda_0) - (\lambda(t)-\lambda_0)(c(t)-c_0) + Q_0 - \lambda_0 c_0. \quad (47)$$

Now we replace $(\lambda(t)-\lambda_0)(c(t)-c_0)$ by its mean value, and this is the approximation. If we take the mean value of both sides in (47) we get

$$E((c(t)-c_0)') = -\operatorname{Cov}(\lambda(t),c(t)) + Q_0 - \lambda_0 c_0.$$

Under general assumptions, see appendix A2, $E(c'(t)) = 0$ and then $\operatorname{Cov}(\lambda(t),c(t)) = Q_0 - \lambda_0 c_0$. Thus we get the approximate relation

$$(c(t)-c_0)' + \lambda_0(c(t)-c_0) = (Q(t) - Q_0) - c_0(\lambda(t) - \lambda_0) \quad (48)$$

which coincides with formula [6] in Gibbs and Slinn (1973, p. 575) if c_0 is replaced by Q_0/λ_0. Since

$$|c_0 - Q_0/\lambda_0| = |\operatorname{Cov}(\lambda(t),c(t))|/\lambda_0 \leq V_{\lambda(0)}\sqrt{\operatorname{Var}(c(0))} \quad (49)$$

it seems reasonable to replace c_0 by Q_0/λ_0 in those cases where the approximation is reasonable. Since we want to avoid "hidden approximations" we shall, however, keep c_0.

Let $f_\lambda(\omega)$ and $f_Q(\omega)$ be the spectral densities for $\lambda(t)$ and $Q(t)$, cf. section 2.

Define the cross covariance $r_{Q,\lambda}(\tau)$ by

$$r_{Q,\lambda}(\tau) = \mathrm{Cov}(Q(0),\lambda(\tau)).$$

From the simultaneous stationarity of $\lambda(t)$ and $Q(t)$ it follows that they are stationarily correlated, i.e., that $\mathrm{Cov}(Q(t),\lambda(t+\tau) = r_{Q,\lambda}(\tau)$. The cross spectral density, see e.g., Grenander and Rosenblatt (1956, p. 47),

$$f_{Q,\lambda}(\omega) = \frac{1}{2\pi}\int_{-\infty}^{\infty} e^{-i\tau\omega} r_{Q,\lambda}(\tau)d\tau$$

is well-defined. Put $Z(t) = (Q(t) - Q_0) - c_0(\lambda(t) - \lambda_0)$. Then

$$r_Z(\tau) = r_Q(\tau) + c_0^2 r_\lambda(\tau) - c_0(r_{Q,\lambda}(\tau) + r_{\lambda,Q}(\tau))$$

and

$$f_Z(\omega) = f_Q(\omega) + c_0^2 f_\lambda(\omega) - 2c_0 \mathrm{Re}\, f_{Q,\lambda}(\omega)$$

where "Re" means "real part of." The function Re $f_{Q,\lambda}(\omega)$ is called the <u>cospectral</u> <u>density</u>.

From (49) and the theory of linear operations on stationary stochastic processes, see e.g., Grenander and Rosenblatt (1956, pp. 49-51), it follows that

$$f_c(\omega) \approx \frac{1}{\lambda_0^2 + \omega^2}(f_Q(\omega) + c_0^2 f_\lambda(\omega) - 2c_0 \mathrm{Re}\, f_{Q,\lambda}(\omega))$$

and thus

$$\mathrm{Var}(c(0)) \approx \int_{-\infty}^{\infty} f_c(\omega)d\omega. \tag{50}$$

The approximate equality is due to the fact that (48) is approximate. It may be noted that the imaginary part of $f_{Q,\lambda}(\omega)$ is an odd function, and thus either $f_{Q,\lambda}(\omega)$ or Re $f_{Q,\lambda}(\omega)$ may be used in (50).

Now we consider the same special case as Gibbs and Slinn (1973). Assume that

$$r_Q(\sigma) = \sigma_Q^2 e^{-A_Q|\tau|}, \quad r_\lambda(\tau) = \sigma_\lambda^2 e^{-A_\lambda|\tau|} \quad \text{and} \quad r_{Q,\lambda}(\tau) = \rho\sigma_Q\sigma_\lambda e^{-A_{Q,\lambda}|\tau|}.$$

Then

$$f_Q(\omega) = \frac{\sigma_Q^2 A_Q}{\pi(A_Q^2 + \omega^2)} \quad \text{and} \quad \int_{-\infty}^{\infty} \frac{1}{\lambda_0^2 + \omega^2} f_Q(\omega)d\omega = \frac{\sigma_Q^2}{\lambda_0} \frac{1}{A_d + \lambda_0}.$$

Thus

$$\mathrm{Var}(c(0)) \approx \frac{\sigma_Q^2}{\lambda_0} \frac{1}{A_Q + \lambda_0} + \frac{c_0^2\sigma_\lambda^2}{\lambda_0} \frac{1}{A_\lambda + \lambda_0} - \frac{2c_0\rho\sigma_Q\sigma_\lambda}{\lambda_0} \frac{1}{A_{Q,\lambda} + \lambda_0} \tag{51}$$

or

$$V_{c(0)} \approx \left(\frac{Q_0}{c_0\lambda_0}\right)^2 \frac{\lambda_0 V_Q^2}{A_Q + \lambda_0} + \frac{\lambda_0 V_\lambda^2}{A_\lambda + \lambda_0} - \left(\frac{Q_0}{c_0\lambda_0}\right)^2 \frac{2\rho\lambda_0 V_Q V_\lambda}{A_{Q,\lambda} + \lambda_0} \tag{52}$$

which coincides with formula (12) in Gibbs and Slinn (1973, p. 575) if c_0 is replaced with Q_0/λ_0.

Up to now we have merely reproduced the derivation in Gibbs and Slinn (1973). We shall now indicate certain situations where the approximation seems to work well.

<u>Case 1</u> (Deterministic sink and random source)

Put $\lambda(t) \equiv \lambda_0$. In this case $c_0 = Q_0/\lambda_0$ and no approximation is involved. Then (47) is reduced to

$$(c(t) - c_0)' + \lambda_0(c(t) - c_0) = (Q(t) - Q_0)$$

and thus

$$\mathrm{Var}(c(0)) = \int_{-\infty}^{\infty} \frac{1}{\lambda_0^2 + \omega^2} f_Q(\omega)d\omega.$$

<u>Case 2</u> (Random sink and deterministic source)

Put $Q(t) = Q_0$. Then (47) is only reduced to

$$(c(t)-c_0)' + \lambda_0(c(t)-c_0) =$$

$$= c_0(\lambda(t)-\lambda_0) - (\lambda(t)-\lambda_0)(c(t)-c_0) + Q_0 - \lambda_0 c_0$$

and the fundamental questions about the approximation remains. We shall consider this case in some detail, since it is technically simpler than the general case. We note that the approximation is reduced to

$$Var(c(0)) \simeq c_0^2 \int_{-\infty}^{\infty} \frac{1}{\lambda_0^2 + \omega^2} f_\lambda(\omega) d\omega$$

and thus

$$V_{c(0)}^2 \simeq \int_{-\infty}^{\infty} \frac{1}{\lambda_0^2 + \omega^2} f_\lambda(\omega) d\omega.$$

Now we consider the case with small fluctuations in $\lambda(t)$. Formally we consider some stationary and non-negative stochastic process $R(t)$ with mean R_0 and put $\lambda(t) = \lambda_0 + a(R(t) - R_0)$. Thus for $a \leq \lambda_0/R_0$, $\lambda(t)$ is non-negative and we have

$$f_\lambda(\omega) - a^2 f_R(\omega).$$

For small values of a we are close to the deterministic case, and and it is reasonable to assume that c_0 is, in general, close to Q_0/λ_0. Furthermore both $c(t)$ and $\lambda(t)$ fluctuate only a little and thus the approximation seems to work in general. Therefore we get

$$Var(c(0)) \simeq \frac{a^2 Q_0^2}{\lambda_0^2} \int_{-\infty}^{\infty} \frac{1}{\lambda_0^2 + \omega^2} f_R(\omega) d\omega.$$

In the special case where $R(t)$ is a two-state Markov process, we can compare this formula with the exact result, and the approximation works. We shall return to this in "case 3" in an example which covers this case.

Now we consider "long-lived particles." Thus we put $\lambda(t) = a R(t)$ and $Q_0 = a\gamma_0$. In this case we have, under certain conditions, that $c_0 \simeq Q_0/\lambda_0 = \gamma_0/R_0$ and we are led to the approximation

$$Var(c(0)) \simeq \frac{\gamma_0^2}{R_0^2} \int_{-\infty}^{\infty} \frac{1}{a^2 R_0^2 + \omega^2} a^2 f_R(\omega) d\omega =$$

$$= \frac{a\gamma_0^2}{R_0^2} \int_{-\infty}^{\infty} \frac{\frac{1}{a}}{R_0^2 + (\frac{\omega}{a})^2} f_R(\omega) d\omega.$$

Since

$$\int_{-\infty}^{\infty} \frac{\frac{1}{a}}{R_0^2 + (\frac{\omega}{a})^2} \, d\omega = \frac{\pi}{R_0} \quad \text{and} \quad \frac{\frac{1}{a}}{R_0^2 + (\frac{\omega}{a})^2} \to 0 \quad \text{for } \omega \neq 0 \text{ as } a \to 0$$

and since $f_R(\omega)$ is bounded and continuous we have

$$\text{Var}(c(0)) \simeq \frac{a\gamma_0^2 2\pi f_R(0)}{2R_0^3}$$

which, since $2\pi f_R(0) = \Gamma_h$, coincides with (33).

Case 3   (Random sink and random source)

Consider first the case with small fluctuations in $\lambda(t)$ and $Q(t)$. Thus we put $\lambda(t) = \lambda_0 + a(R(t) - R_0)$ and $Q(t) = Q_0 + a(\gamma(t) - \gamma_0)$ where $R(t)$ and $\gamma(t)$ are some stationary and non-negative processes with means R_0 and γ_0. For $a \leq \min(\lambda_0/R_0, \, Q_0/\gamma_0)$, both $\lambda(t)$ and $Q(t)$ are non-negative. By the same argument as in "case 2" we assume that $c_0 \simeq Q_0/\lambda_0$. Then we are led to the approximation

$$\text{Var}(c(0)) \simeq a^2 \int_{-\infty}^{\infty} \frac{1}{\lambda_0^2 + \omega^2} \left(f_\gamma(\omega) + \frac{Q_0^2}{\lambda_0^2} f_R(\omega) - 2 \frac{Q_0}{\lambda_0} f_{\gamma,R}(\omega)\right) d\omega.$$

Let $R(t)$ and $\gamma(t)$ be generated by a common two-state Markov process as in the model considered in appendix A4. From section 2, it then follows that

$$f_\gamma(\omega) = \frac{(\gamma_0 - \gamma_d^2)}{\pi p_p \tau_p (A_X^2 + \omega^2)} \quad , \quad f_R(\omega) = \frac{(R_0 - R_d)^2}{\pi p_p \tau_p (A_X^2 + \omega^2)}$$

and

$$f_{\gamma,R}(\omega) = \frac{(\gamma_0 - \gamma_d)(R_0 - R_d)}{\pi p_p \tau_p (A_X^2 + \omega^2)} \quad \text{where} \quad A = \frac{1}{\tau_d} + \frac{1}{\tau_p} \, .$$

It follows from (51) that

$$\mathrm{Var}(c(0)) \simeq \frac{a^2}{\lambda_0 p_p \tau_p (A_X + \lambda_0) A_X} \{(\gamma_0 - \gamma_d)^2 + \frac{Q_0^2}{\lambda^2}(R_0 - R_d)^2 -$$

$$-2\frac{Q_0}{\lambda_0}(\gamma_0 - \gamma_d)(R_0 - R_d)\} = \frac{a^2(\gamma_0 - \gamma_d - Q_0 \frac{R_0 - R_d}{\lambda_0})^2}{\lambda_0 p_p \tau_p (A_X + \lambda_0)} .$$

Since $\lambda_0 p_p \tau_p (A_X + \lambda_0) A_X = \lambda_0 (A_X + \lambda_0) p_p / p_d = \lambda_0 / (p_d \tau_d) + \lambda_0^2 p_p / p_d = (\lambda_0 + \tau_d p_p \lambda_0^2)/(p_d \tau_d)$ it follows from (81) that the approximation works well in this case.

Now we consider "long-lived particles" and put $\lambda(t) = a\, R(t)$ and $Q(t) = a\gamma(t)$. We assume that $f_{\gamma,R}(\omega)$ is bounded and continuous. We further assume, also in this case, that $c_0 \simeq \gamma_0/R_0$. The assumption is supported by (49) and by the considerations later in this section. By the same arguments as used in "case 2" we get

$$\mathrm{Var}(c(0)) \simeq \frac{a}{2R_0}(2\pi f_\gamma(0) + \frac{\gamma_0^2}{R_0^2} 2\pi f_R(0) - \frac{2\gamma_0}{R_0} 2\pi f_{\gamma,R}(0)). \qquad (53)$$

Thus $\mathrm{Var}(c(0)) \propto a$ and since $c_0 \propto 1$ and $E(T) \propto a^{-1}$ we have

$$V_{c(0)} \propto a^{1/2} \propto E(T)^{-1/2}.$$

If $f_{\gamma,R}(0) = 0$, which is the case when the sink and the source are independent, (53) reduces to (39).

Now consider the two-state Markov model again. Then

$$\mathrm{Var}(c(0)) \simeq \frac{a}{R_0 p_p^2 \tau_p A_X^2}(\gamma_0 - \gamma_d - \gamma_0 \frac{R_0 - R_d}{R_0})^2 =$$

$$= \frac{a p_d \tau_d (\gamma_d - R_d \gamma_0 / R_0)^2}{R_0}$$

and, cf. (79), the approximation works well.

We shall now consider approximations of c_0 in the "Gibbs and Slinn sense". Since $r_{\lambda,c}(0) = \mathrm{Cov}(\lambda(t),c(t)) = Q_0 - \lambda_0 c_0$ we have

$$c_0 = \frac{Q_0}{\lambda_0} - \frac{r_{\lambda,c}(0)}{\lambda_0} \tag{54}$$

and thus the problem is to find an expression for $r_{\lambda,c}(0)$. Therefore we multiply (48) with $\lambda(0) - \lambda_0$ and take the mean value of both sides. Then we get $r'_{\lambda,c}(t) + \lambda_0 r_{\lambda,c}(t) \approx r_{\lambda,Q}(t) - c_0 r_\lambda(t)$ which leads to $(i\omega + \lambda_0) f_{\lambda,c}(\omega) \approx f_{\lambda,Q}(\omega) - c_0 f_\lambda(\omega)$ and thus

$$r_{\lambda,c}(0) \approx \int_{-\infty}^{\infty} \frac{1}{\lambda_0 + i\omega}\,(f_{\lambda,Q}(\omega) - c_0 f_\lambda(\omega))\,d\omega =$$

$$= \int_{-\infty}^{\infty} \frac{\lambda_0 - i\omega}{\lambda_0^2 + \omega^2}\,(f_{\lambda,Q}(\omega) - c_0 f_\lambda(\omega))\,d\omega =$$

$$= \lambda_0 \int_{-\infty}^{\infty} \frac{1}{\lambda_0^2 + \omega^2}\,\{\mathrm{Re}\, f_{\lambda,Q}(\omega) + \omega \mathrm{Im}\, f_{\lambda,Q}(\omega)/\lambda_0 - c_0 f_\lambda(\omega)\}\,d\omega =$$

$$= -\lambda_0 \int_{-\infty}^{\infty} \frac{1}{\lambda_0^2 + \omega^2}\,\{c_0 f_\lambda(\omega) - f_{Q,\lambda}(\omega) + \omega \mathrm{Im}\, f_{Q,\lambda}(\omega)/\lambda_0\}\,d\omega$$

where $\mathrm{Im}\, f_{Q,\lambda}(\omega)$ is the imaginary part of $f_{Q,\lambda}(\omega)$. We shall only be interested in cases where $\int \ldots d\omega$ is small, and thus $c_0 \approx Q_0/\lambda_0$ and therefore the c_0 in the integral may be replaced by Q_0/λ_0 and we are led to the approximation

$$c_0 \approx \frac{Q_0}{\lambda_0} + \int_{-\infty}^{\infty} \frac{1}{\lambda_0^2 + \omega^2}\,\{\frac{Q_0}{\lambda_0} f_\lambda(\omega) - f_{Q,\lambda}(\omega) + \omega \mathrm{Im}\, f_{Q,\lambda}(\omega)/\lambda_0\}\,d\omega. \tag{55}$$

Now we consider the same cases as in "case 3." If $\lambda(t) = \lambda_0 + (R(t) - R_0)$ and $Q(t) = Q_0 + (\gamma(t) - \gamma_0)$ we thus get

$$c_0 \simeq \frac{Q_0}{\lambda_0} + a^2 \int_{-\infty}^{\infty} \frac{1}{\lambda^2 + \omega^2} \left\{ \frac{Q_0}{\lambda_0} f_R(\omega) - f_{\gamma,R}(\omega) + \omega \mathrm{Im}\, f_{Q,\lambda}(\omega)/\lambda_0 \right\} d\omega$$

which in the two-state Markov case reduces to

$$c_0 \simeq \frac{Q_0}{\lambda_0} + \frac{p_d \tau_d a^2}{\lambda_0 + \tau_d p_p \lambda_0^2} \left\{ \frac{Q_0}{\lambda_0} (R_0 - R_d)^2 - (\gamma_0 - \gamma_d)(R_0 - R_d) \right\}$$

which coincides with (80).

In the case of "long-lived" particles where $\lambda(t) = a\,R(t)$ and $Q(t) = a\,\gamma(t)$ we have, since $\mathrm{Im}\, f_{Q,\lambda}(0) = 0$

$$c_0 \approx \frac{\gamma_0}{R_0} + \frac{a}{2R_0} \left(\frac{\gamma_0}{R_0} 2\pi f_R(0) - 2\pi f_{\gamma,R}(0) \right). \tag{56}$$

which in the two-state Markov case coincides with (78), and when $f_{\gamma,R}(0) = 0$, with (19).

Mathematical remark

We emphasize that the Gibbs and Slinn approach is a powerful method to heuristically derive approximations, while the results given in appendix A3 are stringent theorems. From a meteorological point of view this might be rather irrelevant, but mathematically, the difference is large. In order to illustrate what me mean, we consider (63) in section 7. If $E(\tilde{T}_d^3) < \infty$, then (63) is in accordance with (52), but if $E(\tilde{T}_d^3) = \infty$ then $\mathrm{Var}(c(0)) = \infty$ for all values of a. The value of $E(\tilde{T}_d^3)$ never appears in the Gibbs and Slinn approach. Although the assumptions given in appendix A3 certainly are not necessary, this illustrates that some kind of assumptions are required.

7 PRECIPITATION SCAVENGING

In this section we adopt a more "practical" approach which partly differ from the rest of the paper. Let as usual $h(t)$ denote the precipitation process and – in case of an intensity model – $R(t)$ denote the precipitation intensity. When nothing else is said we assume that $\Lambda(t) = ah(t)$ whether a is small or not. Recall from section 3 that the value of a ranges from 0.1 - 1. Further, we assume that the source is deterministic and that $Q_0 = a\gamma_0$. Without loss of generality, we further put $\gamma_0 = 1$ which simplifies the formulae.

Let us recall some results from sections 4 and 5. For small values of a, we have

$$E(T) \approx \frac{1}{aR_0} + \frac{\Gamma_h}{2R_0^2} \tag{57}$$

and thus

$$E(c_a(0)) \approx \frac{1}{R_0} + \frac{a\Gamma_h}{2R_0^2} \tag{58}$$

and

$$\mathrm{Var}(c_a(0)) \approx \frac{a\Gamma_h}{2R_0^3}\,. \tag{59}$$

The stringent proofs – see appendix A3 – of (57) and (59) are based on Taylor expansions with respect to a.

Rodhe and Grandell (1981) proposed (57) to be used for $a < 0.2$. This proposal was based on numerical comparisons based on Swedish precipitation data.

Consider an intensity model $R(t)$. We shall always assume that the realizations of $R(t)$ are piecewise constant. This means that we disregard the fluctuations of the precipitation intensity within a precipitation period. Like in the point process case we denote the length of a "typical" dry period by $\tilde{T}_d$ and the length of a "typical" precipitation period and the total amount of precipitation within that period by $(\tilde{T}_p, \tilde{M})$. The precipitation intensity in the period is thus $\tilde{M}/\tilde{T}_p$. Let $\tilde{W}_b$ be the time from the last occurrence of precipitation until 0; i.e., $\tilde{W}_b = \min(t; t \geq 0$ and $R(-t) > 0)$. Thus $\tilde{W}_b = 0$ if $R(0) > 0$ and, cf. section 2,

$$\Pr\{\tilde{W}_b > x\} = \frac{p_d}{\tau_d} \int_x^\infty (1 - F_{\tilde{T}_d}(y))dy \tag{60}$$

and

$$E(\tilde{W}_b^k) = \frac{p_d}{\tau_d} \int_0^\infty x^k (1 - F_{\tilde{T}_d}(y))dy = \frac{p_d E(\tilde{T}_d^{\,k+1})}{\tau_d (k+1)} \tag{61}$$

where $\tau_d = E(\tilde{T}_d)$ and $p_d = \Pr\{R(0) = 0\}$ as usual. All intensity models for $\lambda(t)$ considered in sections 4 and 5 are of this kind. Further, all periods are independent, which makes the interpretation of the word "typical" obvious. If we formally put $\tilde{T}_p = 0$, we get a point process model and the notation completely agree with those used in that case. Whenever we refer to a model for $\Lambda(t)$, we understand, without comments, that the model is applied to $h(t)$.

For $a > 0.2$, Rodhe and Grandell (1981) proposed the generalization of the Markov model, discussed in section 4 just before formula (14). In that case

$$E(c_a(0)) = \frac{1}{R_0} + \frac{a p_d E(\tilde{T}_d^2)}{2\tau_d} \tag{62}$$

and, cf. (83),

$$\mathrm{Var}(c_a(0)) = \frac{a p_d E(\tilde{T}_d^2)}{2\tau_d R_0} + a^2 \{ \frac{p_d E(\tilde{T}_d^3)}{3\tau_d} - (\frac{p_d E(\tilde{T}_d^2)}{2\tau_d})^2 \}. \tag{63}$$

The discussion in Rodhe and Grandell (1981) was only about $E(T)$, and thus only appeals to (62).

One way to understand why (62) works as well as it does is to consider "short-lived" particles; i.e., to consider large values of a. For all a, it holds that

$$c_a(0) = a\tilde{W}_b + c_a(-\tilde{W}_b).$$

For large values of a the concentration ought to be much larger in dry periods than in precipitation periods and we are led to the approximation

$$c_a(0) \approx a\tilde{W}_b \tag{64}$$

and thus to, cf. the mathematical remark in section 5,

$$E(c_a(0)) \approx a\, E(\tilde{W}_b) = \frac{ap_d E(\tilde{T}_d^2)}{2\tau_d} \tag{65}$$

which coincides with the dominating term in (62). Since $c_a(-\tilde{W}_b) > 0$ it follows that $c_a(0) > a\tilde{W}_b$ and $E(c_a(0)) > aE(\tilde{W}_b)$. Thus it seems natural to add some quantity to $aE(\tilde{W}_b)$ in order to get an approximation which works also for "less large" values of a. Since $1/R_0$ is the first approximation for small values of a, that quantity seems to be the natural one to add to $aE(\tilde{W}_b)$, and thus our approximation agrees with (62). Thus we have rediscovered (62) by a reasoning – certainly heuristic – where hardly anything is assumed about R(t). Note that dry and precipitation periods are not assumed to be independent. In the mathematical remark in section 5 we discussed the interpretation of the word "typical". The critical part in our reasoning is that $\lambda(t)$ is assumed to be exactly zero in dry periods. If dry deposition is allowed; i.e., if $\lambda(t) > 0$ also in dry periods, the arguments for (65) breaks down completely.

Consider now $\mathrm{Var}(c(0))$ and put, for notational reasons,

$$v(a) = a^{-1}\mathrm{Var}(c_a(0)).$$

Using the above heuristic reasoning Grandell (1982, p. 254) proposed the approximation

$$v(a) \approx \Gamma_h/(2R_0^3) + a\mathrm{Var}(\tilde{W}_b) = \Gamma_h/(2R_0^3) + a\left\{\frac{p_d E(\tilde{T}_d^3)}{3\tau_d} - \left(\frac{p_d E(\tilde{T}_d^2)}{2\tau_d}\right)^2\right\} \tag{66}$$

and carried through some numerical comparisons, based on Swedish precipitation data, which indicated that (66) works reasonably well.

It is natural to try a "long-life" approximation of $v(a)$, i.e., to try to make a Taylor expansion of $v(a)$ for small values of a, of the form $v(0) + av'(0)$. We have not managed to derive any reasonable expression for $v'(0)$. The figures in Grandell (1982, p. 255) indicate that such an approximation might work only for very small values of a and thus be of limited interest. To be honest, we must admit that the conclusions which can be made from those figures are uncertain and that we would have investigated a "long-life" approximation much more carefully if we had managed to produce one.

We have several times referred to numerical comparisons. Those comparisons are based on precipitation data recorded in Stockholm during summer and winter 1966. Thus the "parameters" R_0, Γ_h, $E(\tilde{W}_b)$ and $Var(\tilde{W}_b)$ which are estimated have to be interpreted as Eulerian. For the year 1966 we have (almost) detailed data; i.e., we have an observation of $R(t)$ for $0 \leq t \leq t_0$, where $t_0 = 4368$ h (half a year). Then R_0, $E(\tilde{W}_b)$ and $Var(\tilde{W}_b)$ are estimated in the "natural" way; i.e., $R_0^* = h(t_0)/t_0$ and $E(\tilde{W}_b^k)^* = \dfrac{p_d^* E(\tilde{T}_d^{k+1})^*}{\tau_d^*(k+1)}$ where

$$E(\tilde{T}_d^k)^* = \frac{1}{N} \sum_{j=1}^{N} \tilde{T}_{d,j}^k \tag{67}$$

and $\tilde{T}_{d,1}, \ldots, \tilde{T}_{d,N}$ are the successive lengths of dry periods. The "parameter" Γ_h is more difficult to estimate, but theoretically it is no problem due to the theory of spectral analysis. For a discussion the practical problems, we refer to Rodhe and Grandell (1981, pp. 374-375).

Generally the precipitation data do not contain such detailed information, but consist of precipitation amounts accumulated over 6, 12 or 24 hours. Let $\{ h_k : k = 1,2,\ldots,n \}$, where $n = t_0/\Delta$, be the total amounts of precipitation during successive time intervals of length Δ(h); i.e.,

$$h_k = \int_{(k-1)\Delta}^{k\Delta} R(x)dx.$$

When $R(t)$ is assumed to be a two-state Markov process, which corresponds to the Markov model with R_p known, estimation problems have been studied by Alexander (1981) and Karr (1984). Under the sole assumption of ergodicity Grandell (1983a) studied estimates of τ_d, τ_p and $E(\tilde{W}_b)$. Roughly speaking it turned out that τ_d and τ_p were difficult to estimate while $E(\tilde{W}_b)$ was simple to estimate.

The use of integrated data does not complicate the estimation of R_0 and Γ_h and therefore we only consider estimation of $E(\tilde{W}_b)$ and $Var(\tilde{W}_b)$. Define

$$\tilde{W}_k(\Delta) = j\Delta \text{ if } h_k=0,\ h_{k-1}=0,\ \ldots,\ h_{k-j+1}=0,\ h_{k-j}>0;$$

i.e., $\tilde{W}_k(\Delta)$ is the length of dry Δ-periods up to time k. Define $\tilde{T}_1(\Delta), \tilde{T}_2(\Delta), \ldots, \tilde{T}_K(\Delta)$ to be the lengths of successive sequences of dry Δ-periods. This means, for example, that if $h_1 = 0$, $h_2 = 0$, $h_3 > 0$, $h_4 = 0$, $h_5 > 0 \ldots$ then $\tilde{T}_1(\Delta) = 2\Delta$, $\tilde{T}_2(\Delta) = \Delta$, ... and $\tilde{W}_1(\Delta) = \Delta$, $\tilde{W}_2(\Delta) = 2\Delta$, $\tilde{W}_3(\Delta) = 0$, $\tilde{W}_4(\Delta) = \Delta$, $\tilde{W}_5(\Delta) = 0 \ldots$. It can always be dicussed how the first $\tilde{W}_k(\Delta)$:s shall be defined. Since generally the length of the time interval $(0,t_0)$ is several months and Δ at most 24 h we shall disregard all problems of such end-effects.

Put $\tilde{W}(\Delta) = \Delta[\tilde{W}_b/\Delta]$, where $[\cdot]$ means integer part, and note that $\tilde{W}(\Delta)$ and $\tilde{W}_k(\Delta)$ have the same distribution. Define $Z(\Delta)$ by

$$\tilde{W}_b = \tilde{W}(\Delta) + Z(\Delta)$$

and thus we have

$$E(\tilde{W}_b) = E(\tilde{W}(\Delta)) + E(Z(\Delta))$$

and

$$Var(\tilde{W}_b) = Var(\tilde{W}(\Delta)) + Var(Z(\Delta)) + 2\,Cov(\tilde{W}(\Delta), Z(\Delta)).$$

The main terms $E(\tilde{W}(\Delta))$ and $Var(\tilde{W}(\Delta))$ are naturally estimated by

$$E(\tilde{W}(\Delta))^* = \frac{1}{n} \sum_{n=1}^{n} \tilde{W}_k(\Delta)$$

and

$$\mathrm{Var}(\tilde{W}(\Delta))^* = \frac{1}{n}\sum_{k=1}^{n} \tilde{W}_k^2(\Delta) - (E(\tilde{W}(\Delta))^*)^2.$$

In order to simplify the numerical computations we put D equal to to the number of dry Δ-periods, i.e., $D = \#\{h_k = 0\}$, and note that

$$\frac{1}{n}\sum_{k=1}^{n} \tilde{W}_k(\Delta) = \frac{1}{2t_0}(\Delta^2 D + \sum_{k=1}^{K} \tilde{T}_k^2(\Delta))$$

and

$$\frac{1}{n}\sum_{k=1}^{n} \tilde{W}_k^2(\Delta) = \frac{1}{6t_0}(\Delta^3 D + 3\Delta \sum_{k=1}^{K} \tilde{T}_k^2(\Delta) + 2\sum_{k=1}^{K} \tilde{T}_k^3(\Delta))$$

In order to estimate the correction terms we assume that $\tilde{T}_d$ is exponentially distributed. Then $\tilde{W}(\Delta)$ and $Z(\Delta)$ are independent conditioned on $\{\tilde{W}_b > 0\}$ and $\tilde{W}(\Delta) = Z(\Delta) = 0$ conditioned on $\{\tilde{W}_b = 0\}$. Under this assumption we have

$$E(Z(\Delta)) = p_d E(Z(\Delta) \mid \tilde{W}_b > 0)$$

$$\mathrm{Var}(Z(\Delta)) = p_d E(Z^2(\Delta) \mid \tilde{W}_b > 0) - (p_d E(Z(\Delta) \mid \tilde{W}_b > 0))^2$$

and

$$\mathrm{Cov}(\tilde{W}(\Delta), Z(\Delta)) = (1 - p_d) E(\tilde{W}(\Delta)) E(Z(\Delta) \mid \tilde{W}_b > 0).$$

Thus we must estimate p_d, $E(Z(\Delta) \mid \tilde{W}_b > 0)$ and $E(Z^2(\Delta) \mid \tilde{W}_b > 0)$. Using the estimates of τ_d and τ_p given by Grandell (1983a), we are led to the estimate

$$p_d^*(\Delta) = \frac{D^2}{n(D-K)} = \frac{\Delta D^2}{t_0(D-K)}. \tag{68}$$

One way to motivate this estimate without using the estimates of τ_d and τ_p is to note that

$$P(\tilde{W}(\Delta) \geq \Delta) = p_d P(\tilde{W}(\Delta) \geq \Delta \mid \tilde{W}_b > 0) = p_d P(\tilde{W}(\Delta) \geq 2\Delta \mid \tilde{W}(\Delta) \geq \Delta).$$

Since it is natural to estimate $P(\tilde{W}(\Delta) \geq \Delta)$ by D/n and to estimate $P(\tilde{W}(\Delta) \geq 2\Delta \mid \tilde{W}(\Delta) \geq \Delta)$ by (D-K)/D, we are again led to (68). Now we proceed as if $Z(\Delta)$, conditioned on $\{\tilde{W}_b > 0\}$, were uniformly distributed on $(0,\Delta)$. Then

$$E(Z(\Delta)|\tilde{W}_b > 0) = \Delta/2 \quad \text{and} \quad E(Z^2(\Delta)|\tilde{W}_b > 0) = \Delta^2/3.$$

Putting all of this together, we are led to the estimates

$$E(\tilde{W}_b)^* = E(\tilde{W}(\Delta))^* + \Delta^2/2$$

and

$$\mathrm{Var}(\tilde{W}_b)^* = \mathrm{Var}(\tilde{W}(\Delta))^* + (1 - p_d^*(\Delta))E(\tilde{W}(\Delta))\Delta + p_d^*(\Delta)\Delta^2/3 - p_d^*(\Delta)^2\Delta^2/4.$$

Using the data mentioned, we compare in Tables 1 and 2 the estimates for different Δ:s. The values for $\Delta = 0$ correspond to the natural estimates based on detailed data.

Δ	$p_d^*(\Delta)$	$E(\tilde{W}(\Delta))^*$	$E(\tilde{W}_b)^*$	$\mathrm{Var}(\tilde{W}(\Delta))^*$	$\mathrm{Var}(\tilde{W}_b)^*$
0	0.836	31.44	31.44	1420	1420
2	0.834	31.07	31.90	1326	1437
6	0.785	28.99	31.34	1387	1429
12	0.706	26.11	30.35	1293	1401
24	0.708	23.21	31.71	1265	1492

TABLE 1: Illustration of estimates. Based on winter data.

Δ	$p_d^*(\Delta)$	$E(\tilde{W}(\Delta))^*$	$E(\tilde{W}_b)^*$	$\mathrm{Var}(\tilde{W}(\Delta))^*$	$\mathrm{Var}(\tilde{W}_b)^*$
0	0.958	80.58	80.58	9500	9500
2	0.955	80.04	81.00	9492	9499
6	0.942	77.98	80.80	9414	9444
12	0.918	75.79	81.30	9497	9585
24	0.870	71.47	81.91	9159	9440

TABLE 2: Illustration of estimates. Based on summer data.

From these tables it seems as if the estimates of $E(\tilde{W}_b)$ and $Var(\tilde{W}_b)$ work well. For some general remarks about why $E(\tilde{W}_b)$ is easier to estimate than τ_d and τ_p we refer to Grandell (1983a, pp. 267 - 268). Those remarks also apply to $Var(\tilde{W}_b)$ and p_d.

The comparisons carried through by Rodhe and Grandell (1981) and Grandell (1982) are based on a mixture of theoretical assumptions and empirical data, and due to this mixture we talk about the <u>empirical model</u>. The idea is to use actual precipitation data and, using (5), to reconstruct the concentration process. Thus the empirical model seems relevant as a reference in the comparisons. On the other hand, because of the use of (5), the comparisons give no information about the general limitations of these kind of models.

We shall (mainly) restrict ourselves to the case where we have integrated data at our disposal. Let c_k be the concentration at the end of the k:th Δ-period. During that period the total amount of precipitation is h_k. The concentrations are recursively calculated according to

$$c_k = c_{k-1}\exp(-ah_k) + a\int_0^{\Delta} \exp(-ah_k t/\Delta)\,dt =$$

$$= \begin{cases} c_{k-1} + \Delta a & \text{if } h_k = 0 \\ c_{k-1}\exp(-ah_k) + \Delta(1 - \exp(-ah_k))/h_k & \text{if } h_k > 0. \end{cases}$$

As a starting value we use $c_0 = a/(aR_0^*) = 1/R_0^*$. All calculations of quantities according to the empirical model are in terms of the reconstructed concentrations $c_1, c_2, \ldots, c_n$. The reconstruction is based on (5) under the assumption that the precipitation intensity is constant during each Δ-period. Put

$$c_0(\text{emp}) = \bar{c} = \frac{1}{n}\sum_{k=1}^{n} c_k$$

and

$$v(a)(\text{emp}) = \frac{a^{-1}}{n}\sum_{k=1}^{n} (c_k - \bar{c})^2,$$

where "emp" stands for "empirical".

We shall compare c_0(emp) with

$$c_0(l.l) = 1/R_0^* + a\Gamma_h^*/(2(R_0^*)^2),$$

see (58), and

$$c_0(s.l) = 1/R_0^* + aE(\tilde{W}_b)^*,$$

see (62) and (65), where "l.l" ("s.l") stands for "long-life" ("short-life"). Further we compare v(a)(emp) with

$$v(a)(s.l) = \Gamma_h^*/(2(R_0^*)^3) + a\mathrm{Var}(\tilde{W}_b)^*,$$

see (66).

In the comparisons we use the same sets of data as Rodhe and Grandell (1981), namely the data from 1966 already used and data from Stockholm for the years 1970 - 72. The last sets of data are daily data; i.e., integrated data with Δ = 24 h. We do, however, omit the data set for the summer 1971 since that was abnormal in the the sense that 55 % of the total amount of precipitation fell in two different weeks separated by three weeks of more dry days. No models seem to work for such a period. In Table 3, some characteristics of the data are given. The values of Γ_h^* are taken from Rodhe and Grandell (1981).

	Winter 1966	Summer 1966	Summer 1970	Winter 1970-71	Winter 1971-72
n	182	182	183	182	183
R_0^*	0.0822	0.0612	0.0508	0.0531	0.0523
Γ_h^*	1.2	1.2	0.7	0.4	0.4
$E(\tilde{W}_b)^*$	31.71	81.91	74.39	34.51	53.37
$\mathrm{Var}(\tilde{W}_b)^*$	1492	9440	9109	1944	4393

TABLE 3: Some characteristics of the data.

As an illustration of the data we give in Figure 4 the daily amounts of precipitation during one year.

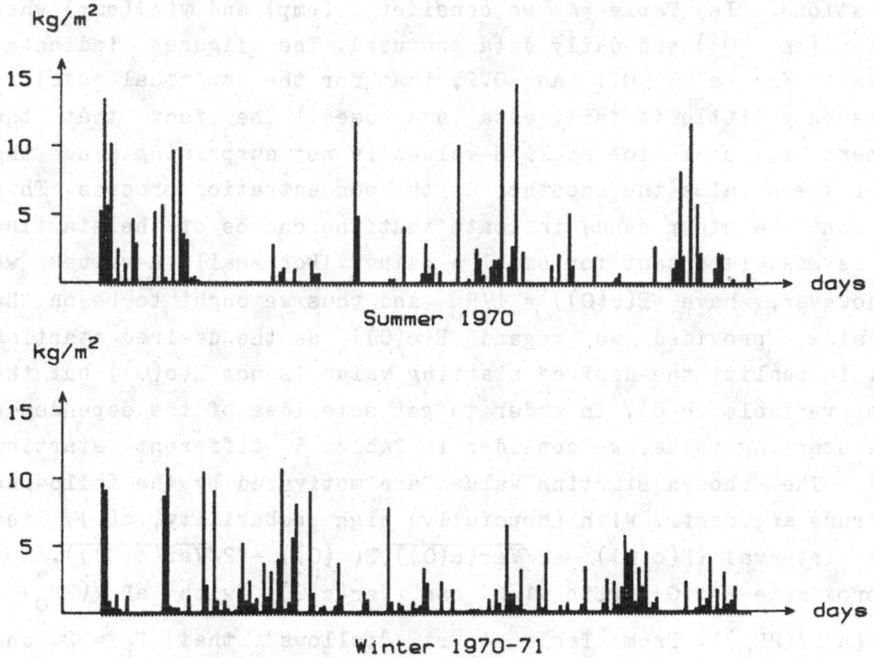

FIGURE 4: Illustration of the daily amounts of precipitation.

We have chosen to consider the a-values 0.1, 0.2 and 1.0. The values a = 0.1 and 1.0 are the "extreme" values in the realistic

	a	Winter 1966		Summer 1966	
		$\Delta = 0$	$\Delta = 24$	$\Delta = 0$	$\Delta = 24$
c_0(emp)	0.1	17.12	17.05	25.53	25.47
v(a)(emp)	0.1	612	609	1525	1545
c_0(emp)	0.2	20.92	20.70	33.91	33.62
v(a)(emp)	0.2	756	757	2358	2386
c_0(emp)	1.0	51.66	43.89	99.52	95.22
v(a)(emp)	1.0	2156	2069	9499	9396

TABLE 4: Comparison for the empirical model between detailed data and daily data.

interval. The value $a = 0.2$ is the "dividing" value between the proposed approximations, and, further, it is a realistic value for applications. In Table 4 we consider $c_0(\text{emp})$ and $v(a)(\text{emp})$ when detailed ($\Delta = 0$) and daily data are used. The figures indicate, at least for $a = 0.1$ and 0.2, that for the empirical model it matters only little if daily data are used. The fact that the agreement is best for small a-values is not surprising since the smaller the a-value the smoother is the concentration process. This does, on the other hand, indicate that the choice of the starting value is most important for small a-values. For small a-values we do, however, have $E(c(0)) \simeq 1/R_0$ and thus we ought to be on the safe side, provided we regard $E(c(0))$ as the desired starting value. In reality the desired starting value is not $E(c(0))$ but the random variable $c(0)$. In order to get some idea of the dependence on the starting value, we consider in Table 5 different starting values. The chosen starting values are motivated by the following very crude arguments. With (hopefully) high probability, $c(0)$ lies in the interval $(E(c(0)) - 2\sqrt{\text{Var}(c(0))}, E(c(0)) + 2\sqrt{\text{Var}(c(0))})$. Now we approximate $E(c(0))$ with $1/R_0$ and $\text{Var}(c(0))$ with $a\Gamma_h/(2R_0^3) = R_0^{-2}\{a\Gamma_h/(2R_0)\}$. From Table 4 it "follows" that $\Gamma_h \simeq 2$ and $R_0 \simeq 0.05$ and "thus" $\text{Var}(c(0)) \simeq R_0^{-2}\{20a\}$ and therefore we consider the starting values 0 (since $c(0) \geq 0$), $1/R_0$ and $(1/R_0)(1 + 2\sqrt{20a})$.

Starting value	Winter 1966	Summer 1966	Summer 1970	Winter 1970-71	Winter 1971-72
			$c_0(\text{emp})$		
0	16.72	24.20	30.58	23.43	25.89
$1/R_0$	17.05	25.47	30.70	23.68	26.33
$(1/R_0)(1 + 2\sqrt{20a})$	17.97	29.07	31.03	24.39	27.58
			$v(a)(\text{emp})$		
0	618	1592	3095	1114	2443
$1/R_0$	609	1545	3040	1075	2325
$(1/R_0)(1 + 2\sqrt{20a})$	717	2623	2972	1034	2411

TABLE 5: Comparison for the empirical model between different starting values when $a = 0.1$.

The figures in Table 5 indicate, possibly with the exception of the summer 1966, that the choice of the starting value is not too important. We have also made comparisons when a = 0.2 and 1.0, and in those cases, as expected, the figures differ less. Consider, as an example, the summer 1966 where v(a)(emp) equals, with increasing starting values 2409, 2386 and 2989 when a = 0.2 and 9420, 9396 and 9584 when a = 1.0.

In Table 6 we compare the approximations with the empirical model.

	a	Winter 1966	Summer 1966	Summer 1970	Winter 1970-71	Winter 1971-72
c_0(emp)	0.1	17.05	25.47	30.70	23.68	26.33
c_0(l.l)	0.1	21.06	32.36	33.25	25.90	26.43
c_0(s.l)	0.1	15.34	24.53	27.13	22.27	24.46
v(a)(emp)	0.1	609	1545	3040	1075	2325
v(a)(s.l)	0.1	1231	3562	3581	1527	1838
c_0(emp)	0.2	20.70	33.62	40.87	28.28	33.70
c_0(l.l)	0.2	29.95	48.39	46.81	32.99	33.74
c_0(s.l)	0.2	18.51	32.72	34.56	25.72	29.79
v(a)(emp)	0.2	757	2386	3918	1183	3261
v(a)(s.l)	0.2	1380	4506	4492	1722	2277
c_0(emp)	1.0	43.89	95.22	99.24	57.51	75.51
c_0(l.l)	1.0	101.04	176.56	155.33	89.65	92.24
c_0(s.l)	1.0	43.88	98.25	94.08	53.33	72.49
v(a)(emp)	1.0	2069	9396	11314	2436	6133
v(a)(s.l)	1.0	2573	12058	11779	3277	5791

TABLE 6: Some comparisons based on the data.

Certainly it is not quite easy to draw definite conclusions from Table 6. A reader interested in numerical comparisons is strongly recomended to consult Rodhe and Grandell (1981) and Grandell (1982) where the comparisons are illustrated by figures. Our general impression is that the short-life approximations are rather good for a $\geq$ 0.2, but it must be kept in mind that their heuristic motivation breaks down completely if dry deposition is allowed. The long-life approximation is theoretically much more satisfying, but the numerical agreement between c_0(emp) and c_0(1.1) is not too impressive. The main reason for the differences is probably, cf. Rodhe and Grandell (1981, p 379), that Γ_h^* has somewhat unpleasant statistical properties. As a consequence Rodhe and Grandell (1981, pp. 379 - 380) recomended use of the long-life approximation only if several years of data are used in the estimation of Γ_h. Using 11 years of precipitation data from Stockholm, they got Γ_h^* = 1.3 (summer period) and 0.6 (winter period).

It is possible that the conclusions and recommendations presented here also have some relevance for other areas with a similar climate. However, no extrapolations can be made to other climatic regimes.

8 THE CONCENTRATION PROCESS

Up to now we have essentially been interested in the <u>random variable</u> $c(0)$ and not in the <u>stochastic process</u> $c(t)$. The reason is, cf. the discussion in section 3, that the models considered describe $c(t)$ while the interesting concentration is $c_E(t)$. In the derivation we have, on the other hand, often used the stationarity of $c(t)$, which is a stochastic process property. Thus it is mathematically highly natural to consider the stochastic process $c(t)$. When the source is deterministic we might consider $c(t)$ in an Eulerian sense, cf. again section 3. In that case it might, also from the point of view of applications, be relevant to consider $c(t)$ as a stochastic process. For purely mathematical reasons we shall, however, sometimes let the source be random.

Put, as before, $r_c(\tau) = \mathrm{Cov}(c(t),c(t+\tau))$ and recall that $r_c(0) = \mathrm{Var}(c((t))$. We shall consider $r_c(\tau))$ in the case where the vector process $(\Lambda(t),q(t))$ has independent increments. This means that the random vectors

$$(\Lambda(t_2) - \Lambda(t_1),q(t_2) - q(t_1)),\ldots,(\Lambda(t_n) - \Lambda(t_{n-1}),q(t_n) - q(t_{n-1}))$$

are independent for any n and any $t_1 < t_2 < \ldots < t_n$. When $\Lambda(t)$ and $q(s)$ are independent it is enough to require that each of them has independent increments. Thus the cases where $\Lambda(t)$ and $q(t)$ are either deterministic or Poisson are included. From the point of view of applications the interesting case is probably when $\Lambda(t)$ is Poisson and $q(t)$ is deterministic. Further the point process model considered in appendix A4, where $\Lambda(t)$ and $q(s)$ may be dependent, is included.

Since $\Lambda(t)$ has stationary and independent increments which implies that $\Lambda(1)$ is infinitely divisible, it follows, cf. Feller (1971, p. 450), that

$$G(t) = \exp\{-t\psi(1)\}$$

for some constant $\psi(1)$. In the deterministic case $\psi(1) = \lambda_0$ and in the Poisson case, cf. (15), $\psi(1) = (1 - \phi(1))/\tau_d$. It follows from (5) that

$$c_0 = \int_{-\infty}^{t} E(e^{-(\Lambda(t) - \Lambda(s))})E(dq(s)) = Q_0 \int_{-\infty}^{t} G(t-s)ds = Q_0/\psi(1)$$

since $\Lambda(t) - \Lambda(s)$ and $dq(s)$ are independent. For $\tau > 0$ we have

$$c(\tau) = c(0)e^{-\Lambda(t)} + \int_0^t e^{-(\Lambda(\tau) - \Lambda(s))}dq(s)$$

and thus

$$r_c(\tau) = -c_0^2 + G(\tau)E(c^2(0)) + c_0Q_0 \int_0^{\tau} G(\tau-s)ds =$$

$$= -c_0^2 + G(\tau)E(c^2(0)) + c_0Q_0(1 - G(\tau))/\psi(1) = G(\tau)r_c(0)$$

since $c(0)$, $dq(s)$ and $\Lambda(\tau) - \Lambda(s)$ are independent. Thus we have

$$r_c(\tau) = G(|\tau|)r_c(0)$$

for all τ. This result is due to Baker et al. (1979, pp. 47 and 50) in the two cases when $\Lambda(t)$ or $q(t)$ is Poisson and the other is deterministic.

Grandell (1982, p.246) computed $r_c(\tau)$ in the case when the sink is a Markov model with $\lambda_d = 0$ and the source is deterministic. The result is rather complicated and will not be reproduced here. More interesting would be to consider $r_c(\tau)$ when the sink is a S. R. renewal model, but we have not managed to perform the calculations in that case.

Now we consider "long-lived particles"; i.e., when $\Lambda(t) = ah(t)$, $\lambda(t) = aR(t)$, $q(t) = ag(t)$ and $Q(t) = a\gamma(t)$ as before. As a simple example we assume that the source is deterministic and that $h(t)$ is a S. R. Markov model. Then we have, cf. (24),

$$r_c(\tau) = r_c(0)\ \exp\{-a|\tau|R_0/(1 + aR_0\tau_d)\}$$

which illustrates the fact, mentioned in section 7, that $c(t)$ becomes smoother when a becomes smaller.

Under general assumptions we have $c_0 \approx \gamma_0/R_0$ and $r_c(0) \propto a$ for small values of a and thus it seems natural to introduce the normalized and "time-transformed" process

$$\bar{c}_a(t) = (c(t/a) - \gamma_0/R_0)/\sqrt{a}.$$

Then we have

$$r_{\bar{c}_a}(\tau) = r_c(\tau/a)/a.$$

Define further

$$\bar{h}_a(t) = (h(t/a) - R_0 t/a)\cdot\sqrt{a}$$

and note that $\mathrm{Var}(\bar{h}_a(t)) = a\mathrm{Var}(h(t/a)) \to t\Gamma_h$ as $a \to 0$.

Consider now a variant of the Gibbs and Slinn approximation method and assume that the sink and the source are described by intensity models. Define $\bar{R}_a(t)$ and $\bar{\gamma}_a(t)$ by

$$\bar{R}_a(t) = \bar{h}_a'(t) = (R(t/a) - R_0)/\sqrt{a}$$

and

$$\bar{\gamma}_a(t) = (\gamma(t/a) - \gamma_0)/\sqrt{a}.$$

Then we have, cf. (47) and (48),

$$\bar{c}_a'(t) + R_0\bar{c}_a(t) = \bar{\gamma}_a(t) - (\gamma_0/R_0)\bar{R}_a(t) - \sqrt{a}\ \bar{c}_a(t)\bar{R}_a(t)$$

and the approximate relation

$$\bar{c}_a'(t) + R_0\bar{c}_a(t) = \bar{\gamma}_a(t) - (\gamma_0/R_0)\bar{R}_a(t).$$

Thus, cf. the approximate form of the spectral density $f_c(\omega)$, we have

$$f_{\bar{c}_a}(\omega) \approx \frac{1}{R_0^2 + \omega^2}\left(f_{\bar{\gamma}_a}(\omega) + (\gamma_0^2/R_0^2)f_{\bar{R}_a}(\omega) - 2(\gamma_0/R_0)\mathrm{Re}\ f_{\bar{\gamma}_a,\bar{R}_a}(\omega)\right).$$

Since $r_{\bar{\gamma}_a}(\tau) = r_\gamma(\tau/a)/a$ we have $f_{\bar{\gamma}_a}(\omega) = f_\gamma(a\omega)$ and similarly $f_{\bar{R}_a}(\omega) = f_R(a\omega)$ and $f_{\bar{\gamma}_a,\bar{R}_a}(\omega) = f_{\gamma,R}(a\omega)$.

Thus we get, for small values of a,

$$f_{\bar{c}_a}(\omega) \approx \frac{1}{R_0^2 + \omega^2}(f_\gamma(0) + (\gamma_0^2/R_0^2)f_R(0) - 2(\gamma_0/R_0)f_{\gamma,R}(0))$$

and

$$r_{\bar{c}_a}(\tau) \approx e^{-R_0|\tau|}(2\pi f_\gamma(0) + (\gamma_0^2/R_0^2)2\pi f_R(0) - (\gamma_0/R_0)4\pi f_{\gamma,R}(0))/(2R_0)$$

which, for $\tau = 0$, reduces to (53).

Now we restrict ourselves to the case with deterministic source. Recalling that $\Gamma_h = 2\pi f_R(0)$ we are led to the approximation

$$r_{\bar{c}_a}(\tau) \approx e^{-R_0|\tau|}\gamma_0^2\Gamma_h/(2R_0^3).$$

Instead of discussing approximations of the covariance function further we shall consider approximations of the normalized concentration $\bar{c}_a(t)$ itself. Assume that h(t) is "approximately normal" for large values of t which implies, cf. (21), that

$$\bar{h}_a(t) \overset{d}{\approx} \sqrt{t\Gamma_h} \cdot W$$

for all values of t as a is small. Since normality is perserved under linear operations it "follows" that $\bar{R}_a(t)$ is approximately normal – derivation is a linear operation – and "thus" $\bar{c}_a(t)$ is approrximately normal since $\bar{c}_a(t)$ is approximately linearly related with $\bar{R}_a(t)$. Let X(t) be a normal process with mean zero and covariance function

$$r_X(\tau) = e^{-R_0|\tau|}\gamma_0^2\Gamma_h/(2R_0^3)$$

which means that X(t) is an Ornstein-Uhlenbeck process. Thus it seems reasonable to believe that $\bar{c}_a(t)$ behaves like X(t) for small values of a.

We strongly admit that this heuristic "derivation" is very far from stringent. A different – somewhat more convincing but more complicated – heuristic reasoning is given by Grandell (1982, pp. 248 - 249). The fact that $\bar{c}_a(t)$ may be approximated by $X(t)$ is nevertheless true and the precise formulation is given in appendix A3.

Consider now the random variable $c(0)$. Then

$$\bar{c}_a(0) \overset{d}{\to} \gamma_0\{\Gamma_h/(2R_0^3)\}^{1/2}W \text{ as } a \to 0$$

and thus we are led to the <u>approximation</u>

$$c(0) \overset{d}{\approx} \gamma_0(R_0^{-1} + \{\Gamma_h/(2R_0^3)\}^{1/2}W)$$

for small values of a. The underlying normalization is based on the approximations $c_0 \approx \gamma_0/R_0$ and $Var(c(0)) \approx a\gamma_0^2\Gamma_h/(2R_0^3)$. If we base the normalization on the approximations, cf. (58) and (66),

$$c_0 \approx \gamma_0(R_0^{-1} + a\Gamma_h/(2R_0^2)) \text{ and } Var(c(0)) \approx a\gamma_0^2\Gamma_h/(2R_0^3) + a^2\gamma_0^2Var(\tilde{W}_b)$$

we are led to the <u>modified approximation</u>

$$c(0) \overset{d}{\approx} \gamma_0(\{R_0^{-1} + a\Gamma_h/(2R_0^2)\} + \{a\Gamma_h/(2R_0^3) + a^2Var(\tilde{W}_b)\}^{1/2}W).$$

Certainly the approximation may by modified in a variety of ways. To be precise we may consider any normalizing functions $A(a)$ and $B(a)$ such that, cf. Feller (1971, p 253),

$$(A(a) - \gamma_0/R_0)/\sqrt{a} \to 0 \text{ and } B^2(a)/a \to \gamma_0^2\Gamma_h/(2R_0^3) \text{ as } a \to 0$$

since then $(c(0) - A(a))/B(a) \overset{d}{\to} W$ and we are led to

$$c(0) \overset{d}{\approx} A(a) + B(a)W.$$

If $A(a) = E(c(0))$ and $B^2(a) = Var(c(0))$ we talk about the <u>natural approximation</u>.

In order to get some idea about how good the approximations are we consider the case when $h(t)$ is described by a S. R. Markov model. In that case $\Gamma_h = 2R_0^2\tau_d$ and $Var(\tilde{W}_b) = \tau_d^2$ and thus the approximation is reduced to

$$c(0) \overset{d}{\cong} \gamma_0(R_0^{-1} + \sqrt{a\tau_d/R_0}\, W)$$

and the modified approximation to

$$c(0) \overset{d}{\cong} \gamma_0(R_0^{-1} + a\tau_d + \{a\tau_d(R_0^{-1} + a\tau_d)\}^{1/2} W).$$

In this case c(0) is gamma distributed, but before discussing that we shall give some basic facts about the gamma distribution.

A non-negative random variable X is said to be $\Gamma(\alpha,\beta)$-distributed if its density function $f_X(x)$ is given by

$$f_X(x) = \frac{x^{\alpha-1}\exp\{-x/\beta\}}{\beta^{\alpha}\Gamma(\alpha)}, \quad x \geq 0,$$

where $\Gamma(\cdot)$ is the gamma-function. This implies that $E(X) = \alpha\beta$, $Var(X) = \alpha\beta^2$ and that its Laplace-transform $\hat{f}_X(u)$ is given by

$$\hat{f}_X(u) = (1 + \beta u)^{-\alpha}.$$

It is well known that

$$X \overset{d}{\cong} \beta(\alpha + \sqrt{\alpha}\cdot W)$$

for large values of α. For reference reasons we call this the <u>gamma approximation</u>.

In the mathematical remark it is shown that c(0) is $\Gamma(\alpha,\beta)$-distributed with

$$\alpha = 1 + (aR_0\tau_d)^{-1} \text{ and } \beta = a\gamma_0\tau_d.$$

Thus the modified approximation, the natural approximation and the gamma approximation coincide in this case. Although we shall use the distribution of c(0) only for comparisons, such knowledge has an interest in itself. In appendix A4, cf. (101) and (104), we have derived the distribution of c(0) for two related models with random source. In both cases the approximate normality holds for small values of a.

In order to compare the approximate distributions with the gamma distribution we consider the <u>ε-point</u> $c_\varepsilon(0)$ of the concentration, defined by $\Pr\{c(0) > c_\varepsilon(0)\} = \varepsilon$. Thus $c_\varepsilon(0)$ is the value which $c(0)$ exceeds with probability ε. Let $X = A + BW$ be any approximation, i.e. $c(0) \overset{d}{\approx} X$. Then $X_\varepsilon = A + BW_\varepsilon$ is the corresponding approximation of $c_\varepsilon(0)$. For fixed ε it is natural to consider the relative error $(X_\varepsilon - c_\varepsilon(0))/c_\varepsilon(0)$. In practical applications it may be equally or more interesting to consider the <u>actual ε</u> defined by $\Pr\{c(0) > X_\varepsilon\}$. The actual ε is thus a measure of "how often" the concentration exceeds the approximate ε-point.

In Table 7 the approximation and the modified approximation are compared with the exact distribution. In the comparisons we put $\tau_d = 20$ and $R_0 = 0.1$ which are rather realistic values in connection with precipitation scavenging. Further, like in section 7, we put $\gamma_0 = 1$.

		Approximation			Modified approximation			Exact
ε	a	ε-point	rel. error	act. ε	ε-point	rel. error	act. ε	ε-point
10%	0.2	18.11	-25%	25%	23.59	-1.8%	10.7%	24.03
10%	0.1	15.73	-15%	20%	18.28	-1.5%	10.7%	18.55
5%	0.2	20.40	-27%	18%	26.31	-6.5%	6.9%	28.13
5%	0.1	17.36	-17%	14%	20.06	-4.6%	6.6%	21.03
1%	0.2	24.71	-33%	9%	31.41	-15.0%	2.8%	36.95
1%	0.1	20.40	-22%	6%	23.40	-10.8%	2.5%	26.22

TABLE 7: Comparison for the S.R. Markov model between approximate and exact distributions.

Roughly speaking the approximation seems to work badly compared with the modified approximation. Both approximations work, quite naturally, better for a = 0.1 than for a = 0.2. They also work better for larger ε-values. This is also quite natural, since the approximations are essentially based on the central limit theorem which generally works best in the "probable area".

Consider now the empirical model discussed in section 7. Let $c_1, c_2, \ldots, c_n$ be the empirical concentrations based on daily data. They are illustrated in Figure 5 for two of our sets of data. The reader is advised to compare with Figure 4 in section 7 where the corresponding daily amounts of precipitation are drawn.

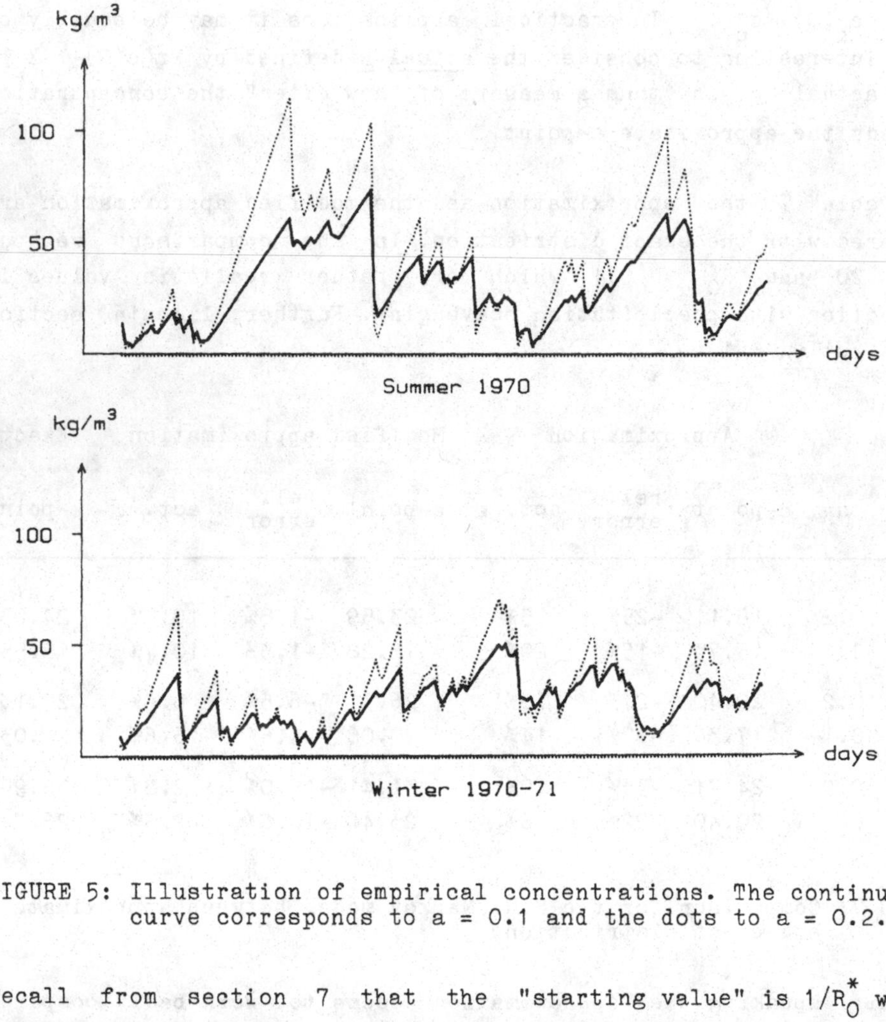

FIGURE 5: Illustration of empirical concentrations. The continuous curve corresponds to a = 0.1 and the dots to a = 0.2.

Recall from section 7 that the "starting value" is $1/R_0^*$ which explains why the concentration is low in the beginning of a period. Within the accuracy of Figure 5 the effect of the starting value, cf. Table 5, is at most about 10 days for the summer 1970 and about 20 days for the winter 1970-71. The question whether the empirical concentrations look like Ornstein-Uhlenbeck processes is left to the reader.

Define the empirical distribution function $F_c^*(x)$ by

$$F_c^*(x) = \#\{c_k \leq x\}/n$$

Thus $F_c^*(x)$ jumps by $1/n$ at each point c_k, and, with the notation used in section 7, $c_0(\text{emp})$ and $av(a)(\text{emp})$ are its mean and variance respectively. In the comparison for the empirial model F_c^* plays the rôle of the exact distribution function. Let <u>emp. ε-point</u> be the ε-point with respect to F_c^*. Since F_c^* increases by jumps the emp. ε-point always coincides with one of the c_k:s. Let the <u>e.n.a. ε-point</u>, where e.n.a. stands for empirical natural approximate, be $c_0(\text{emp}) + \sqrt{av(a)(\text{emp})}\cdot W_\varepsilon$. The relative error and the actual ε are defined as in the comparison for the S.R. Markov model. In Table 8 we compare the approximate and the empirical distributions.

It is natural to compare Table 8 with the modified approximation in Table 7. Roughly speaking the approximations work about equally well in the S.R. Markov model and in the empirical model for $\varepsilon = 10\%$ and $\varepsilon = 5\%$ and much better in the empirical model for $\varepsilon = 1\%$. Although conclusions must be drawn with care for these comparisons our general impression is that the normal approximation of $c(0)$ works rather well for $a < 0.2$ provided the normalizing functions are close to the mean and the standard deviation of $c(0)$.

It might be surprising that the approximation seems to work so relatively good in the empirical model for $\varepsilon = 1\%$. In this case, since $n = 182$ or 183, the emp. ε-point equals the second largest among $c_1,\ldots,c_n$. Define emp. c_{max} by

$$\text{emp. } c_{max} = \max(c_1, c_2, \ldots, c_n).$$

From the construction of the c_k:s it follows that

$$\text{emp. ε-point} \leq \text{emp. } c_{max} \leq 24a + \text{emp. ε-point}$$

and thus the emp. ε-point and the emp. c_{max} are close to each other for small values of a. It might be illustrative to compare the emp. ε-points, for $\varepsilon = 1\%$, in Table 8 with the emp. c_{max}:s in Table 10. Mathematically there is a great difference between "approximations of the maximal concentration" and "approximations of ε-points," and

	ε	a	Winter 1966	Summer 1966	Summer 1970	Winter 1970-71	Winter 1971-72
emp. ε-point	10%	0.1	28.10	42.30	55.93	37.05	46.61
e.n.a. ε-point	10%	0.1	27.05	41.40	53.04	36.97	45.87
rel. error	10%	0.1	-3.7%	-2.1%	-5.2%	-0.2%	-1.6%
actual ε	10%	0.1	12.1%	11.0%	14.2%	10.4%	10.4%
emp. ε-point	10%	0.2	38.82	66.63	81.31	50.25	63.30
e.n.a. ε-point	10%	0.2	36.47	61.61	76.75	47.99	66.43
rel. error	10%	0.2	-6.0%	-7.5%	-5.6%	-4.5%	4.9%
actual ε	10%	0.2	13.2%	13.2%	12.6%	13.2%	8.2%
emp. ε-point	5%	0.1	31.77	47.04	60.19	40.75	56.91
e.n.a. ε-point	5%	0.1	29.88	45.92	59.38	40.73	51.41
rel. error	5%	0.1	-5.9%	-2.4%	-1.3%	-0.0%	-9.7%
actual ε	5%	0.1	6.6%	6.0%	6.0%	5.5%	6.6%
emp. ε-point	5%	0.2	45.90	78.82	93.87	56.61	89.01
e.n.a. ε-point	5%	0.2	40.94	69.55	86.92	53.58	75.70
rel. error	5%	0.2	-10.8%	-11.8%	-7.4%	-5.4%	-14.9%
actual ε	5%	0.2	8.8%	9.3%	7.1%	6.6%	6.6%
emp. ε-point	1%	0.1	35.78	53.80	71.12	49.50	76.11
e.n.a. ε-point	1%	0.1	35.20	54.39	71.26	47.79	61.80
rel. error	1%	0.1	-1.6%	1.1%	0.2%	-3.4%	-18.8%
actual ε	1%	0.1	1.1%	0.5%	0.5%	1.1%	3.8%
emp. ε-point	1%	0.2	50.45	90.80	110.11	67.77	127.41
e.n.a. ε-point	1%	0.2	49.32	84.43	105.99	64.06	93.10
rel. error	1%	0.2	-2.2%	-7.0%	-3.7%	-5.5%	-26.9%
actual ε	1%	0.2	1.6%	2.7%	1.1%	2.2%	4.9%

TABLE 8: Comparison between approximate and empirical distributions.

we believe that the behaviour of the emp. ε-points, for ε = 1%, is more related to "approximations of the maximal concentration" then to "approximations of ε-points."

Now we consider approximations of the maximal concentration. The main reason is certainly their intrinsic interest and not their possible relation to the ε-points in the empirical model.

For any stochastic process $\eta(t)$ we denote its maximal value on the interval $[0,T]$ by $M(\eta,T)$; i.e.,

$$M(\eta,T) = \sup_{0 \leq t \leq T} \eta(t).$$

Our aim is to to consider approximations of $M(c,T)$ for small values of a when T is large. Since we have found that the choice of the normalizing functions is very important, we consider the normalized process

$$c_a(t) = (c(t/a) - A(a))/B(a)$$

instead of $\bar{c}_a(t)$. It is obvious that $c_a(t)$ may be approximated by an Ornstein-Uhlenbeck process $Y(t)$ with mean zero and covariance function

$$r_Y(\tau) = e^{-R_0|\tau|}$$

under the same general conditions as holds for the approximation of $\bar{c}_a(t)$. Under those conditions it follows from appendix A4 that

$$M(c_a,S) \overset{d}{\to} M(Y,S) \text{ as } a \to 0$$

for any fixed value of S. Since

$$M(c_a,S) = (M(c,S/a) - A(a))/B(a)$$

we are led to the approximation

$$M(c,T) \overset{d}{\cong} A(a) + B(a)M(Y,aT)$$

for small values of a and large values of T.

The main theoretical problem in this approximation is that — to our knowledge — the exact distribution of $M(Y,S)$ can not be expressed in a tractable way, cf. Blake and Lindsey (1973, p. 299). We therefore have to rely on an approximation of $M(Y,S)$, see Leadbetter et al. (1983, p. 237), which holds for large values of S. Thus not only T but also aT must be large. This is not without problems since we believe that the larger T the smaller a must be in order to make the approximation $M(c,T) \stackrel{d}{=} A(a) + B(a)M(Y,aT)$ good. Thus the requirements of small a and large aT might be somewhat contradictory. A further problem is that the approximation depends on the unit of the "time-scale" of $Y(t)$.

The unit of the time-scale of $c(t)$ is h (hours) and thus the "time-unit" of $c_a(t)$ is m^2h/kg which hardly is a very "natural time-scale." Obviously the "time-unit" of $Y(t)$ and $c_a(t)$ is the same. Let us therefore define a class Y^δ of "time-transformed" versions of Y defined by

$$Y^\delta(t) = Y(t/\delta).$$

It is easy to realize that $M(Y,S) = M(Y^\delta,\delta S)$ and that Y^δ is an Ornstein-Uhlenbeck process with

$$r_{Y^\delta}(\tau) = e^{-(R_0/\delta)|\tau|}.$$

Now we consider the approximation of $M(Y,S)$. Let Z be a random variable with distribution function

$$F_Z(z) = \exp(-e^{-z}), \quad -\infty < z < \infty.$$

Put

$$\alpha^\delta(S) = (2\log(\delta S))^{1/2}$$

$$\beta^\delta(S) = \alpha^\delta(S) + (1/2)(\log\log(\delta S) + 2\log(R_0/\delta) - \log(\pi))/\alpha^\delta(S).$$

Then it follows from Leadbetter et al. (1983, p. 237) that

$$\alpha^\delta(S)(M(Y,S) - \beta^\delta(S)) \xrightarrow{d} Z \quad \text{as } S \to \infty.$$

This limit theorem, which holds for any choice of δ, leads to the approximation

$$M(Y,S) \overset{d}{\approx} \beta^{\delta}(S) + Z/\alpha^{\delta}(S)$$

which, on the other hand, depends on δ. This problem is, of course, not new and in Leadbetter et al. (1983, p. 293) it is proposed that δ shall be chosen such that "time" is counted in a "natural" way. It seems natural to require that the "time-scale" of Y^{δ} shall be independent of the time-unit of $c(t)$ and of the unit of the concentration. Thus the unit of δ ought to be kg/m^2h since then the "time-scale" of Y^{δ} is dimensionless. This is obtained if $\delta = \theta R_0$ for some dimensionless constant θ. We have no good arguments for the choice of θ. The naive desire to get simple approximations leads to $\theta = 1$ or $1/\sqrt{\pi}$. The two choices yield very similar approximations and we have chosen $\theta = 1$. Thus we use the approximation

$$M(c,T) \overset{d}{\approx} A(a) + B(a)(\beta^{R_0}(aT) + Z/\alpha^{R_0}(aT))$$

for small values of a and (very) large values of T.

Consider now $M(c,T)$ in the case when $h(t)$ is described by a S. R. Markov model with – as before – $\gamma_0 = 1$, $\tau_d = 20$ and $R_0 = 0.1$. Put $T = 4368$ (182 times 24) which implies that $M(c,T)$ is the maximal concentration during half a year. Unfourtunately we have not managed to derive the distribution of $M(c,T)$ analytically, and therefore we have simulated the distribution. The simulation has been done on a Zenith Z-100 microcomputer with the aid of the random number generator included in the Z-BASIC interpretor. For each of the values $a = 0.1$ and 0.2 we have simulated $c(t)$, $0 \leq t \leq T$, 10 000 times. Each simulation of $c(t)$ requires the generation of about 400 random numbers and thus the total study about eight million random numbers. Numerical experiments indicate that this is within the capacity of the random number generator.

Let $F_M(x)$ denote the distribution function of $M(c,T)$ and let $F_M^*(x)$ denote the corresponding empirical distribution function obtained by simulation. Let $M(c,T)_\varepsilon$ and $M^*(c,T)_\varepsilon$ be the ε-points with respect to $F_M(x)$ and $F_M^*(x)$ respectively. It follows from Cramér (1945, p. 369) that $M^*(c,T)_\varepsilon$ is approximately normally distributed with mean $M(c,T)_\varepsilon$ and variance $\varepsilon(1-\varepsilon)/(nf_M^2(M(c,T)_\varepsilon))$ provided $F_M(x)$ has a continuously differentiable density $f_M(x)$ in some neighbourhood of $x = M(c,T)_\varepsilon$. The empirical density function $f_M^*(x)$ is defined by

$$f_M^*(x) = F_M^*([x]+1) - F_M^*([x])$$

where — as before — $[\cdot]$ means integer part. Thus

$$M^*(c,T)_\varepsilon \pm 2(\varepsilon(1-\varepsilon)/n)^{1/2}/\, f_M^*(M^*(c,T)_\varepsilon)$$

is an approximate 95% confidence interval for $M(c,T)_\varepsilon$.

In Table 9 approximate ε-points are compared with the "exact" ones for $\varepsilon = 0.75$, 0.5 and 0.25 which correspond to the lower quartile, the median and the upper quartile. For each ε-value we have also given the actual ε like in Table 7. The "exact" values are $M^*(c,T)_\varepsilon$ and the intervals are the approximate 95% confidence intervals given above. In the approximations we have used $A(a) = E(c(0))$ and $B(a) = \sqrt{Var(c(0))}$.

		Approximation		"Exact"
ε	a	ε-point	act. ε	ε-point
75%	0.1	25.0	69%	24.5 ± 0.1
50%	0.1	26.3	56%	26.8 ± 0.1
25%	0.1	27.8	40%	29.6 ± 0.1
75%	0.2	36.0	75%	36.0 ± 0.2
50%	0.2	37.8	69%	40.0 ± 0.2
25%	0.2	39.9	50%	44.7 ± 0.2

TABLE 9: Comparison for the S.R. Markov model between approximate and "exact" distributions of the maximal concentration during half a year.

As a complement to Table 9 we have drawn the approximate and the "exact" densities in Figure 6.

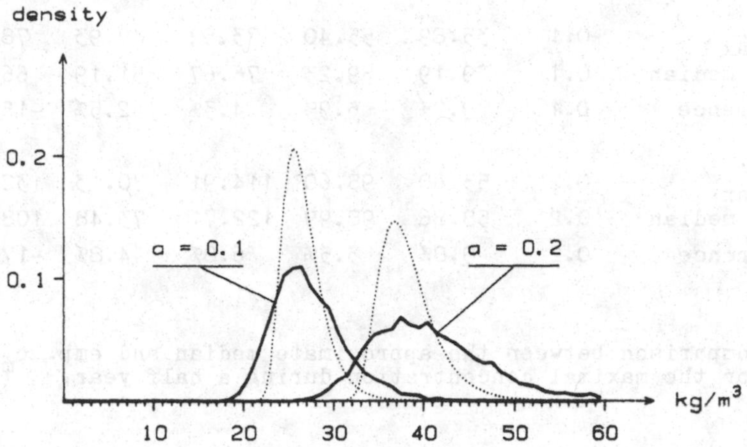

FIGURE 6: Comparison for the S.R. Markov model between approximate and "exact" densities of the maximal concentration during half a year. The continuous curve represents the "exact" density $f_M^*(x)$ and the dots the approximate density.

Table 9 and Figure 6 indicate that the exact and the approximate distributions are "located" close to each other but that the exact distribution is more "spread" than the approximate one. The only conclusion we dare to draw is that the approximate median seems to work as an approximation of the "location" of the maximal concentration.

Consider now the empirical model. In this case we use the approximation

$$M(c,T) \overset{d}{\approx} c_0(\text{emp}) + \sqrt{av(a)(\text{emp})}(\beta_{R_0^*}(aT) + Z/\alpha_{R_0^*}(aT))$$

and the approximate median is with regard to that approximation. In Table 10 we compare the emp. c_{max}:s with the approximate medians. The relative differences are defined by

relative difference = (appr. median - emp. c_{max})/emp. c_{max}.

	a	Winter 1966	Summer 1966	Summer 1970	Winter 1970-71	Winter 1971-72
emp. c_{max}	0.1	35.89	55.40	73.52	49.93	78.51
approximate median	0.1	39.19	59.23	76.67	51.19	66.73
rel. difference	0.1	9.2%	6.9%	4.3%	2.5%	-15.0%
emp. c_{max}	0.2	53.60	95.60	114.91	70.13	132.21
approximate median	0.2	58.86	98.99	122.71	73.48	108.64
rel. difference	0.2	9.8%	3.5%	6.8%	4.8%	-17.8%

TABLE 10: Comparison between the approximate median and emp. c_{max} for the maximal concentration during a half year.

Although any definite conclusions may not be drawn from Table 10, it does – to our opinion – support the use of the approximate median as an approximation of the "location" of the maximal concentration or as an "estimate" of the maximal concentration.

In order to illustrate the approximation for different choices of T we consider again the S. R. Markov model. In Table 11 we compare the approximate ε-point with the "exact" ones in the case a = 0.1. The "exact" values are based on 2500 simulations of c(t). The periods are the T-values in months where each month is taken to 30 days. The relative errors are defined as in Table 7 and the approximate 95% confidence intervals are defined as in Table 9. The "25%-75%"-point means the interquartile range $M(c,T)_{25\%} - M(c,T)_{75\%}$ which is a measure of the dispersion of the distribution.

Table 11 indicates that the approximate median is close to the exact one for periods of 3 to 6 months. The fact that the approximation seems less good for both smaller and longer periods is in accordance with our previous discussion. Due to climatic reasons it is anyhow not advisable to assume stationarity for periods longer than about 6 months. Thus the indicated periods are satisfying from the point of view of our applications.

ε	Period (in months)	Approximate ε-point	"Exact" ε-point	Relative error
75%	1	20.4	17.9 ± 0.2	
50%	1	22.1	20.4 ± 0.2	8.1%
25%	1	24.2	23.6 ± 0.3	
"25%-75%"	1	3.9	5.7 ± 0.3	
75%	3	23.5	21.8 ± 0.2	
50%	3	24.8	24.4 ± 0.2	2.0%
25%	3	26.6	27.4 ± 0.3	
"25%-75%"	3	3.1	5.5 ± 0.3	
75%	6	25.0	24.5 ± 0.2	
50%	6	26.3	27.0 ± 0.2	-2.6%
25%	6	27.8	29.8 ± 0.2	
"25%-75%"	6	2.8	5.3 ± 0.2	
75%	12	26.4	26.8 ± 0.2	
50%	12	27.5	29.1 ± 0.2	-5.6%
25%	12	29.0	31.8 ± 0.2	
"25%-75%"	12	2.6	5.0 ± 0.2	
75%	24	27.6	29.2 ± 0.2	
50%	24	28.7	31.3 ± 0.2	-8.6%
25%	24	30.0	34.0 ± 0.3	
"25%-75%"	24	2.4	4.8 ± 0.2	

TABLE 11: Comparison for the S.R. Markov model with $a = 0.1$ between approximate and "exact" distributions of the maximal concentration during periods of different lengths.

It may be noted that the "location" of the distributions increases slowly with T, which is typical for distributions of extremes. The approximate interquartile ranges decrease slowly to zero. We believe that the exact interquartile ranges, on the other hand, tend to a positive constant. The reason for our belief is that

$$\max_{1 \le k \le n} X_k - \log(n) - (\alpha - 1)\log\log(n) + \log(\Gamma(\alpha)) \overset{d}{\to} \beta Z$$

when $n \to \infty$ if $X_1, \ldots, X_n$ are independent and $\Gamma(\alpha,\beta)$-distributed. In this case the interquartile ranges tend to $\beta(Z_{25\%} - Z_{75\%}) = 1.573\beta$.

Mathematical remark

We now consider the distribution of $c(t)$ when $h(t)$ is described by a S.R. Markov model. In this case $c(t)$ is a "pseudo-Poisson process with linear increments" and the distribution of $c(t)$ can be derived, cf. Feller (1971, pp. 324 - 326), by "Markov-methods". We shall, however, give a somewhat more direct derivation.

Let $\tilde{M}$ be "the amount of precipitation" at a "typical shower", let $\tilde{T}_d$ be the time to the next "shower" and let c_d be the concentration just before that "shower". The random variables c_d, $\tilde{M}$ and $\tilde{T}_d$ are thus independent. The value of the the concentration just before the next "shower" is thus

$$c_d e^{-a\tilde{M}} + a\gamma_0 \tilde{T}_d.$$

If we instead of a "typical shower" choose the "shower" previous to time $t = 0$ it follows that $c(0) \overset{d}{=} c_d$ since $\tilde{W}_b \overset{d}{=} \tilde{T}_d$ when $\tilde{T}_d$ is exponentially distributed. Thus we have the relation

$$c(0) \overset{d}{=} c(0)e^{a\tilde{M}} + a\gamma_0 \tilde{T}_d.$$

Let $\phi_{c(0)}(u)$ be the Laplace transform of $c(0)$; i.e, $\phi_{c(0)}(u) = E(\exp\{-uc(0)\})$. Thus we have

$$\phi_{c(0)}(u) = E(\phi_{c(0)}(ue^{-a\tilde{M}}))\phi_{\tilde{T}_d}(a\gamma_0 u).$$

Since $\tilde{M}$ and $\tilde{T}_d$ are exponentially distributed with means $R_0\tau_d$ and τ_d respectively we get

$$\phi_{c(0)}(u)(1 + a\gamma_0\tau_d u) = (1/(R_0\tau_d)) \int_0^\infty \phi_{c(0)}(ue^{-ax})e^{-x/(R_0\tau_d)} dx =$$

$$= (1/(aR_0\tau_d)) \int_0^u \phi_{c(0)}(y)\, y^{(1/(aR_0\tau_d))-1}\, u^{-1/(aR_0\tau_d)}\, dy$$

or

$$\phi_{c(0)}(u)aR_0\tau_d(1 + a\gamma_0\tau_d u)u^{1/(aR_0\tau_d)} = \int_0^u \phi_{c(0)}(y)\, y^{(1/(aR_0\tau_d))-1} dy.$$

Differentiation and obvious simplifications yield

$$\phi'_{c(0)}(u)(1 + a\gamma_0\tau_d u) = -\phi_{c(0)}(u)a\gamma_0\tau_d\{1 + (1/(aR_0\tau_d))\}$$

or

$$\frac{d \log\phi_{c(0)}(u)}{du} = - \{1 + (1/(aR_0\tau_d))\}a\gamma_0\tau_d/(1 + a\gamma_0\tau_d u).$$

Since $\phi_{c(0)}(0) = 1$ we get

$$\log\phi_{c(0)}(u) = - \{1 + (1/(aR_0\tau_d))\}\log(1 + a\gamma_0\tau_d u)$$

and thus c(0) is $\Gamma(\alpha,\beta)$-distributed with

$$\alpha = 1 + (aR_0\tau_d)^{-1} \text{ and } \beta = a\gamma_0\tau_d.$$

A1 INEQUALITIES FOR THE MEAN CONCENTRATION

We shall now prove the two inequalities mentioned in section 4. In the proofs we rely on Jensen's inequality for convex functions. A function f is called convex if all its cords lie above or on the graph of f. In analytical terms this means that

$$pf(x_1) + (1-p)f(x_2) \geq f(px_1 + (1-p)x_2) \quad \text{for } 0 \leq p \leq 1.$$

Let X be any random variable, such that E(X) is well-defined, and f a convex function. Jensen's inequality states that

$$E(f(X)) \geq f(E(X)).$$

Consider now $G(t) = E(e^{-\Lambda(t)})$. Since the function $f(x) = e^{-x}$ is convex, we immediately get

$$G(t) \geq e^{-E(\Lambda(t))} = e^{-\lambda_0 t}.$$

Consider now c_0 in the case where the sink and the source may be dependent, and assume that the source is described by an intensity model. Then

$$c(0) = \int_{-\infty}^{0} e^{\Lambda(x)} Q(x)dx.$$

Since $f(x) = -\log x$ is convex we have

$$-E(\log(e^{\Lambda(x)}Q(x))) = -x\lambda_0 - E(\log Q(x)) \geq -\log((Ee^{\Lambda(x)}Q(x)))$$

and thus

$$E(e^{\Lambda(x)} Q(x)) \geq e^{x\lambda_0} e^{E(\log Q(x))} = e^{x\lambda_0} e^{E(\log Q(0))}.$$

This implies

$$c_0 \geq \int_{-\infty}^{0} e^{x\lambda_0} e^{E(\log Q(0))} dx = \exp(E(\log Q(0)))/ \lambda_0.$$

A2 CONDITIONS FOR $E(c'(t)) = 0$

In sections 4 and 6 we used the fact that $E(c'(t)) = 0$ under general conditions.

We have

$$c'(t) = \lim_{\Delta \downarrow 0} \frac{c(t+\Delta) - c(t)}{\Delta}$$

and since

$$E\left(\frac{c(t+\Delta) - c(t)}{\Delta}\right) = 0,$$

provided that $E(c(t)) < \infty$, we only have to show that

$$\lim_{\Delta \downarrow 0} E\left(\frac{c(t+\Delta) - c(t)}{\Delta}\right) = E\left(\lim_{\Delta \downarrow 0} \frac{c(t+\Delta) - c(t)}{\Delta}\right). \tag{69}$$

If there exists a non-negative random variable X with $E(X) < \infty$ and such that

$$\sup_{0<\Delta<\Delta_0} \left| \frac{c(t+\Delta) - c(t)}{\Delta} \right| < X \quad \text{for some} \quad \Delta_0 > 0$$

then (69) follows by dominated convergence. By stationarity it is enough to consider $t = 0$. From (5) it follows that

$$c(\Delta) = c(0)e^{-\Lambda(\Delta)} + \int_0^\Delta e^{-(\Lambda(\Delta)-\Lambda(x))} Q(x)dx$$

and thus

$$\frac{c(\Delta) - c(0)}{\Delta} \leq \frac{1}{\Delta} c(0) \left|1 - e^{-\Lambda(\Delta)}\right| + \frac{1}{\Delta} \int_0^\Delta e^{-(\Lambda(\Delta)-\Lambda(x)} Q(x)dx \leq$$

$$\leq c(0) \frac{\Lambda(\Delta)}{\Delta} + \frac{q(\Delta)}{\Delta} < c(0) \sup_{0\leq x\leq \Delta_0} \lambda(x) + \sup_{0\leq x\leq \Delta_0} Q(x) \tag{70}$$

If we put X equal to the last term in (70) we realize that (69) holds if

$$E\left(\sup_{0\le x\le\Delta_0} Q(x)\right) < \infty \quad \text{for some } \Delta_0 > 0$$

and if

$$E\left(c(0)\sup_{0\le x\le\Delta_0} \lambda(x)\right) < \infty \quad \text{for some } \lambda_0 > 0. \tag{71}$$

In the application in section 4 we have $\lambda(x) \le \lambda_d$ and thus (71) holds if $E(c(0)) < \infty$. For the application in section 6 we may note that

$$E\left(c(0)\sup_{0\le x\le\Delta_0} \lambda(x)^2\right) \le E(c^2(0))\, E\left(\sup_{0\le x\le\Delta_0} \lambda^2(x)\right)$$

and thus (71) holds if $E(c^2(0)) < \infty$ and $E\left(\sup_{0\le x\le\Delta_0} \lambda^2(x)\right) < \infty$.

A3 APPROXIMATIONS FOR "LONG-LIVED" PARTICLES

Consider $h(t)$ as given and put $\Lambda(t) = a\, h(t)$. Put $R_0 = E(h(1))$ and $\Gamma_h = \lim_{t\to\infty} Var(h(t))/t$. Let T_a denote the residence time and put $G_a(t) = Pr\{T_a > t\}$. We shall now state the precise formulations of the approximations holding when $a \to 0$. In order to do this we need some technical assumptions. The parameter α is always choosen such that $1 < \alpha \leq 2$.

<u>Assumption 1(α)</u>. There exists a constant v_α such that

$$|Var((h(t)) - \Gamma_h t| \leq v_\alpha t^{2-\alpha} \quad \text{for } t \geq 1.$$

<u>Assumption 2(α)</u>. If $3/2 < \alpha \leq 2$ there exists a constant D_α such that

$$|E\{(h(t) - R_0 t)^3\}| \leq D_\alpha t^{3-\alpha} \quad \text{for } t \geq 1.$$

<u>Assumption 3(β)</u>. There exists a constant B_β such that

$$E\{|h(t) - R_0 t|^{2\beta}\} \leq B_\beta t^\beta \quad \text{for } t \geq 1.$$

<u>Assumption 4</u>. There exists constants $K > 1$ and $C > 0$ such that

$$G_a(t/a) \leq K\, e^{-Ct}$$

for all $t \geq 0$ and all $a_0 <$ some a.

These conditions, which may look rather horrible, are probably not too restrictive in practice. For some technical remarks about them we refer to Grandell (1980, pp. 32-33).

The following results are shown by Grandell (1980, p. 33).

1. If for some α , $1 < \alpha \leq 2$, assumptions $1(\alpha)$, $2(\alpha)$ and $3(\alpha)$ hold, then, for fixed $t > 0$,

$$G_a(t/a) = e^{-R_0 t}\left(1 + \frac{at\Gamma_h}{2}\right) + O(a^\alpha).$$

As usual $O(a^\alpha)$ means that $a^{-\alpha} O(a^\alpha)$ remains bounded as $a \to 0$.

2. If for some α, $1 < \alpha \leq 2$, assumptions $1(\alpha)$, $2(\alpha)$, $3(\beta)$ and 4 hold with $\beta > \alpha$, then

$$E(T_a) = \frac{1}{aR_0} + \frac{\Gamma_h}{2R_0^2} + O(a^{\alpha-1}).$$

It is also shown by Grandell (1980) that any α can occur and that we can not hope to get general results for $\alpha > 2$ unless the approximation contains more terms. It is further shown that an alternative approximation is given by

$$G_a(t/a) = \exp(-R_0 t + at\Gamma_h/2) + O(a^\alpha).$$

Now we consider the concentration process with deterministic source strength $Q_0 = a\gamma_0$ where γ_0 will be considered as a constant. The following result is shown by Grandell (1983b, p. 148).

3. If assumptions 1(2), $3(\beta)$ and 4 hold with $\beta > 3/2$, then

$$\mathrm{Var}(c_{Q_0}(t)) = \frac{a\gamma_0^2\Gamma_h}{2R_0^3} + O(a^{3/2}).$$

From this it follows that

$$V^2_{C_{Q_0}}(0) = a\Gamma_h/(2R_0) + O(a^{3/2}) = \left(E(T_a) - \frac{1}{aR_0}\right)/E(T_a) + O(a^{3/2}).$$

Now we consider the case where also the source is random. Let $g(t)$ be a non-decreasing process with stationary increments such that

$$E(g(1)) = \gamma_0$$

and, cf. (26),

$$\mathrm{Cov}(dg(x),dg(y)) = r_g dx\, d\delta(y-x) + r_\gamma(y-x)dx\, dy.$$

Assumption 5.

$$\int_{-\infty}^{\infty} (1 + |\tau|^{1/2})\, |r_\gamma(\tau)|\, d\tau < \infty.$$

Assumption 5 implies that $\Gamma_g = \lim_{t\to\infty} \mathrm{Var}(g(t))/t$ exists and equals $r_g + \int_{-\infty}^{\infty} r_\gamma(\tau)d\tau$.

Assume that h(s) and g(t) are independent and put $q(t) = ag(t)$.

4. If assumptions 1(2), 3(β), 4 and 5 hold with $\beta > 3/2$, then

$$\mathrm{Var}(c(t)) = \frac{a}{2R_0}\left(\frac{\gamma_0^2\Gamma_h}{R_0^2} + \Gamma_g\right) + O(a^{3/2}).$$

It follows from (27) that

$$\mathrm{Var}(c(0)) = \mathrm{Var}(c_{Q_0}(0)) + a^2 r_q E(T_{2a}) + B$$

where

$$B = 2a^2 \int_{-\infty}^{0}\int_{-\infty}^{0} E(e^{a(h(x) + h(x+y))})r_\gamma(y)\, dy\, dx.$$

Under the assumptions we have $E(T_a) = 1/(aR_0) + O(1)$ and thus

$$a^2 r_g E(T_{2a}) = \frac{ar_g}{2R_o} + O(a^2)$$

and thus it remains to prove that

$$B = \frac{a}{2R_0} \int_{-\infty}^{\infty} r_\gamma(y)dy + O(a^{3/2}). \tag{72}$$

Put $X(t) = h(t) - R_0 t$. Define, for $0 < \delta \leq 1$, the remainder term $H_\delta(x)$ by

$$e^x = 1 - H_\delta(x).$$

Then we have

$$|H_\delta(x)| \le |x|^\delta (1 + e^x).$$

(Obviously H_δ does not depend on δ, but the "δ" indicates which δ is to be used in the above inequality.)

With these notation we have

$$B = 2a^2 \int_{-\infty}^{0} \int_{-\infty}^{0} e^{aR_0(2x+y)} E(e^{a(X(x)+X(x+y))}) r_\gamma(y)\,dy\,dx =$$

$$= \frac{a}{R_0} \int_{-\infty}^{0} e^{R_0 y} r_\gamma(y)\,dy + 2a^2 \int_{-\infty}^{0} \int_{-\infty}^{0} e^{aR_0(2x+y)} E(H_1(a(X(x)+X(x+y)))) r_\gamma(y)\,dy\,dx =$$

$$= (I) + (II).$$

Now we have

$$(I) = \frac{a}{R_0} \int_{-\infty}^{0} (1 + H_{1/2}(aR_0 y)) r_\gamma(y)\,dy = \frac{a}{R_0} \int_{-\infty}^{0} r_\gamma(y)\,dy +$$

$$+ \frac{a}{R_0} \int_{-\infty}^{0} H_{1/2}(aR_0 y) r_\gamma(y)\,dy.$$

It follows from assumption 5 that

$$\left|a \int_{-\infty}^{0} H_{1/2}(aR_0 y) r_\gamma(y)\,dy\right| \le a^{3/2} \int_{-\infty}^{0} |y|^{1/2}(1 + e^{aR_0 y}) r_\gamma(y)\,dy \le \text{const.}a^{3/2}$$

and it remains to consider (II).

Schwartz´ inequality yields

$$E|H_1(a(X(x) + X(x+y)))| \le$$

$$\le a(E((X(x) + X(x+y))^2))^{1/2}(E((1 + e^{a(X(x)+X(x+y))})^2))^{1/2}.$$

Since x,y $\leq$ 0, it follows from assumption 3(1), that

$$E((X(x) + X(x+y))^2) \leq \text{const.}\ (1 - 2x - y).$$

It follows from assumption 4 that

$$E(1 + e^{a(X(x) + X(x+y))})^2 \leq 2(1 + e^{-2aR_0(2x+y)} E\, e^{2aX(x)}\, e^{2aX(x+y)}) \leq$$

$$\leq 2(1 + e^{-2aR_0(2x+y)} G_{4a}(-x)^{1/2}\, G_{4a}(-x-y)^{1/2} \leq \text{const.}(1 + e^{2a(C-R_0)(2x+y)})$$

for $a < a_0/4$.

Thus, since $C \leq R_0$, it follows from assumption 5 that

$$\text{(II)} \leq \text{const.}a^3 \int_{-\infty}^{0}\int_{-\infty}^{0} (1 + |x|^{1/2} + |y|^{1/2}) e^{aC(2x+y)} r_\gamma(y)dy\, dx \leq$$

$$\leq \text{const.}a^3 \int_{-\infty}^{0} (1 + |x|)^{1/2}\, e^{a2Cx} dx \leq \text{const.}a^{3/2}.$$

Now we return to the case with deterministic source strength $a\gamma_0$ and consider an approximation of the concentration process. The approximation is based on the theory of weak convergence of probability measures. Standard references well suited for our applications are Billingsley (1968) and Lindvall (1973).

Let D be the space of functions on $(-\infty,\infty)$ that are right-continuous and have left-hand limits. Endowed with the Skorohod J_1 topology D is a Polish space; i.e., separable and metrizable with a complete metric. A stochastic process $X = \{X(t)\ ;\ -\infty < t < \infty\}$ is said to <u>be in D</u> if all its realizations are in D. The distribution of X is a probability measure on D. Let $X, X_1, X_2, \ldots$ be processes in D. We say that X_n converges in distribution to X, and we write $X_n \xrightarrow{d} X$, if $E(f(X_n)) \to E(f(X))$ for all bounded and continuous real-valued functions f on D. Let T_X be those t-values for which $\Pr\{X(t) = X(t-)\} = 1$. Convergence in distribution of X_n

to X implies, for example, that $X_n(t) \xrightarrow{d} X(t)$ for any fixed $t \in T_X$ and that $\sup_{t_1 \le t \le t_2} X_n(t) \xrightarrow{d} \sup_{t_1 \le t \le t_2} X(t)$ for any t_1 and $t_2 \in T_X$.

Let the processes $\bar{c}_a$ and $\bar{h}_a$ be defined as in section 8. From their definitions it follows that they may be considered as processes in D. Let further W be a standard Wiener process, i.e., $W(0) = 0$, $W(t)$ has independent and normally distributed increments such that $E(W(t) - W(s)) = 0$ and $\mathrm{Var}(W(t) - W(s)) = t - s$ for $t > s$ and its realizations are continuous.

Assumption 6.

$$\bar{h}_a \xrightarrow{d} \sqrt{\Gamma_h} \cdot W \text{ as } a \to 0.$$

Let X be a stationary normal process with $E(X(t)) = 0$ and

$$r_X(\tau) = \frac{\gamma_0^2 \Gamma_h}{2R_0^3} e^{-R_0|\tau|};$$

i.e., X is an Ornstein-Uhlenbeck process.

The following result is due to Grandell (1982, pp. 250 - 251).

5. If assumptions 1(3/2), 4 and 6 hold, then

$$\bar{c}_a \xrightarrow{d} X \text{ as } a \to 0.$$

A4 MODELS WITH DEPENDENT SINK AND SOURCE

Assume that "the environment" met by an air parcel is described by a stationary process $X(t)$ which completely determines the sink intensity and the source strength, in such a way that

$$\lambda(t) = \tilde{\lambda}(X(t)) \quad \text{and} \quad Q(t) = \tilde{Q}(X(t))$$

where $\tilde{\lambda}(x)$ and $\tilde{Q}(x)$ are deterministic functions. This approach may be realistic if the variation is induced by the movement of the air parcel.

Let us first illustrate the extremal – but certainly completely unrealistic – case where $\tilde{Q}(x) = \tilde{c}\,\tilde{\lambda}(x)$ for some constant $\tilde{c}$. Then (5) reduces to

$$c(t) = \tilde{c} \int_{-\infty}^{t} e^{-(\Lambda(t)-\Lambda(s))}\lambda(s)ds = \tilde{c}$$

and thus the sink and source compensate each other in such a way that the concentration becomes deterministic.

Assume now that $X(t)$ is a two-state Markov process taking the values p and d. (The unnatural choice of p and d will get its notational explanation later.) Let τ_p and τ_d be the mean length of a p-period and a d-period respectively and p_p and p_d the probabilities for such periods. For details we refer to section 2. Put $\lambda_p = \tilde{\lambda}(p)$, $\lambda_d = \tilde{\lambda}(d)$, $Q_p = \tilde{Q}(p)$ and $Q_d = \tilde{Q}(d)$. Thus, both $\lambda(t)$ and $Q(t)$ are two-state Markov processes. We shall always assume that $\lambda_d \leq \lambda_p$. The sink process does notationally coincide with the sink process used by Rodhe and Grandell (1972), where d meant "dry" and p meant "precipitation". This coincidence is comfortable, and is the explanation for the unnatural choice. Since

$$E(c(0)) = \int_{-\infty}^{0} E(e^{\Lambda(s)}Q(s))ds$$

we consider

$$E(e^{\Lambda(s)}Q(s)) = p_d E(e^{\Lambda(s)}Q(s)|X(s) = d) + p_p E(e^{\Lambda(s)}Q(s)|X(s) = p) =$$

$$= p_d\, Q_d\, G_d(-s) + p_p\, Q_p\, G_p(-s),$$

where $G_d(t)$ is the probability that a particle which enters the atmosphere in a d-period will have a residence time longer that t. Such probabilities were calculated by Rodhe and Grandell (1972). We further define

$$E_d(T) = \int_0^{\infty} G_d(t)dt,$$

which is the mean residence time for particles entering the atmosphere in a d-period. In the corresponding way we define $G_p(t)$ and $E_p(T)$. Thus we have

$$E(c(0)) = \int_{-\infty}^{0} (p_d Q_d G_d(-s) + p_p Q_p G_p(-s))ds = p_d Q_d E_d(T) + p_p Q_p E_p(T).$$

From Rodhe and Grandell (1972, p. 448) it thus follows that

$$c_0 = \frac{\tau_d Q_0 + \tau_p Q_0 + \tau_d \tau_p (p_d \lambda_p Q_d + p_p \lambda_d Q_p)}{\tau_d \lambda_d + \tau_p \lambda_p + \tau_d \tau_p \lambda_d \lambda_p}, \qquad (73)$$

where $Q_0 = p_d Q_d + p_p Q_p$. In the same way, $\lambda_0 = p_d \lambda_d + p_p \lambda_p$, and we get

$$c_0 = \frac{Q_0}{\lambda_0} + \frac{p_d \tau_d (\lambda_0 - \lambda_d)(Q_d - Q_0 \lambda_d / \lambda_0)}{\lambda_0 (1 + \tau_d p_p \lambda_0) + p_p (\lambda_0 - \lambda_d)(\tau_d \lambda_d - \tau_p \lambda_0)}. \qquad (74)$$

Consider λ_0, Q_0, τ_d and τ_p as fixed. For $\lambda_d = \lambda_0$ we have no variation in sink intensity, and thus we are in the classical case, and consequently we have $c_0 = Q_0/\lambda_0$. For $\lambda_d \neq \lambda_0$ it is seen that c_0 increases linearly in Q_d. The very special case with $Q_d = Q_0 \lambda_d / \lambda_0$ corresponds to a deterministic concentration process with $c(t) = Q_0/\lambda_0$. The case $Q_d = Q_0$ corresponds to a deterministic source strength and thus we have $c_0 = Q_0\, E(T)$. Thus we have

$$c_0 \le Q_0/\lambda_0 \qquad \text{if} \qquad 0 \le Q_d \le Q_0\lambda_d/\lambda_0$$

$$Q_0/\lambda_0 \le c_0 \le Q_0\, E(T) \qquad \text{if} \qquad Q_0\lambda_d/\lambda_0 \le Q_d \le Q_0$$

$$Q_0\, E(T) \le c_0 \qquad \text{if} \qquad Q_0 \le Q_d .$$

In Rodhe and Grandell (1972) it was natural to consider $\lambda_d = 0$. In this case, (73) is reduced to

$$c_0 = \frac{Q_0}{\lambda_0} + p_d\tau_d Q_d. \tag{75}$$

Now we consider the calculation of $Var(c(0))$. Following Grandell (1982, p. 246) we get

$$E(c^2(0)) = E \int_{-\infty}^{0}\int_{-\infty}^{0} e^{\Lambda(x)+\Lambda(y)}\, Q(x)\, Q(y)\, dx =$$

$$= 2 \int_{-\infty}^{0}\int_{-\infty}^{x} E(e^{2\Lambda(x)}\, Q(x)\, e^{\Lambda(y)-\Lambda(x)}\, Q(y))\, dx\, dy.$$

From the Markov property it follows that

$$E(c^2(0))=$$

$$= 2 \int_{-\infty}^{0}\int_{-\infty}^{x} p_d Q_d E(e^{2\Lambda(x)}|\lambda(x) = \lambda_d)E(Q(y)e^{\Lambda(y)-\Lambda(x)}|\lambda(x) = \lambda_d) +$$

$$+ p_p Q_p E(e^{2\Lambda(x)}|\lambda(x) = \lambda_p)E(Q(y)e^{\Lambda(y)-\Lambda(x)}|\lambda(x) = \lambda_p)dydx =$$

$$= 2p_d Q_d E_d(T^{(2)}) \int_0^{\infty} E(Q(0)e^{-\Lambda(t)}|\lambda(t) = \lambda_d)dt +$$

$$+ 2p_p Q_p E_p(T^{(2)}) \int_0^{\infty} E(Q(0)e^{-\Lambda(t)}|\lambda(t) = \lambda_p)dt$$

where $T^{(2)}$ is the residence time when $\lambda(t)$ is replaced by $2\lambda(t)$.

Since

$$E(Q(0)e^{-\Lambda(t)} \mid \lambda(t) = \lambda_d) = E(Q(t)e^{-\Lambda(t)} \mid \lambda(0) = \lambda_d),$$

see, e.g., Kelly (1979, pp. 6-7), and since

$$Q(t) = [\lambda_p Q_d - \lambda_d Q_p + \lambda(t)(Q_p - Q_d)]/(\lambda_p - \lambda_d),$$

we have

$$E(c^2(0)) = \frac{2}{\lambda_p - \lambda_d}\{ p_d Q_d E_d(T^{(2)})[(\lambda_p Q_d - \lambda_d Q_p)E_d(T) + Q_p - Q_d] +$$

$$+ p_p Q_p E_p(T^{(2)}) [(\lambda_p Q_d - \lambda_d Q_p)E_d(T) + Q_p - Q_d]\}.$$

From this we get

$$\mathrm{Var}(c(0)) = \frac{2\lambda_0(Q_d - Q_0\lambda_d/\lambda_0)}{\lambda_0 - \lambda_d}\{ p_d Q_d E_d(T^{(2)})E_d(T) +$$

$$+ p_p Q_p E_p(T^{(2)})E_p(T) - c_0{}^{(2)}/\lambda_0\} + 2\frac{Q_0}{\lambda_0} c_0{}^{(2)} - c_0^2 \quad (76)$$

where $c_0^{(2)}$ is the mean concentration when $\lambda(t)$ is replaced by $2\lambda(t)$. Thus we have expressed $\mathrm{Var}(c(0))$ in known quantities. In the case where $\lambda_d = 0$, this reduces to

$$\mathrm{Var}(c(0)) = Q_d^2\tau_d p_d(\tau_d + \lambda_0^{-1} + p_p\tau_d) \quad (77)$$

The rather complicated formulae (74) and (76) are illustrated in Figure 7 where the relative variance are drawn for the case $Q_0 = \lambda_0 = \tau_d + \tau_p = 1$ and $p_p = 0.5$. The choice $Q_0 = 1$ is irrelevant since $V^2{}_{c(0)}$ is considered and $\tau_d + \tau_p = 1$ is merely a normalization of the time unit. The choices $\lambda_0 = 1$ and $p_d = 0.5$ is not meant to be realistic, but the purpose of the figure is only illustrative. Recall that $Q_d = \lambda_d$ corresponds to a deterministic concentration and that $Q_d = 1$ corresponds to a deterministic source.

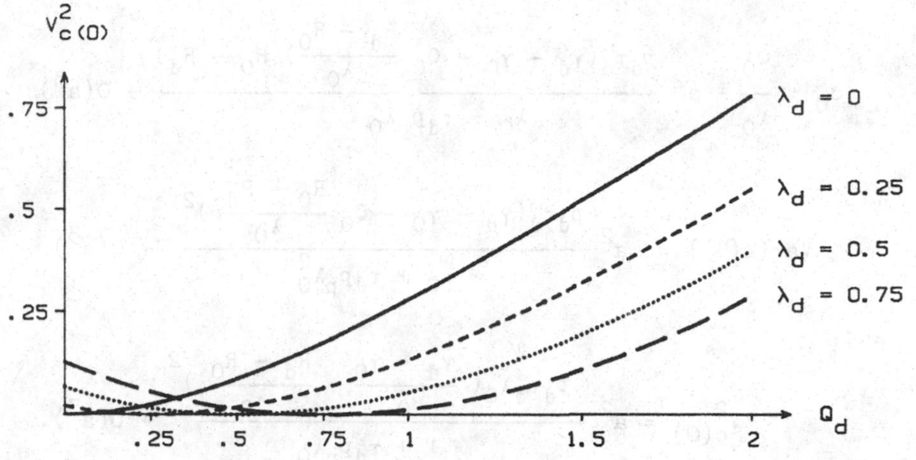

FIGURE 7: Illustration of the relative variance when $p_d = 0.5$.

Consider now "long-lived" paricles. Thus we assume that $\lambda(t)=aR(t)$ and that $Q(t) = a\gamma(t)$ where $R(t)$ and $\gamma(t)$ correspond to $Q(t)$ and $\lambda(t)$ in the above formulae. The quantities R_d, R_0, γ_d and γ_0 have their obvious meaning. Then we have

$$c_0 = \frac{\gamma_0}{R_0} + a\,\frac{p_d\tau_d(\gamma_d - \frac{R_d}{R_0}\gamma_0)(R_0 - R_d)}{R_0} + O(a^2) \qquad (78)$$

$$\mathrm{Var}(c(0)) = \frac{ap_d\tau_d(\gamma_d - \frac{R_d}{R_0}\gamma_0)^2}{R_0} + O(a^2) \qquad (79)$$

and

$$V^2_{c(0)} = ap_d\tau_d R_0\left(\frac{\gamma_d}{\gamma_0} - \frac{R_d}{R_0}\right)^2 + O(a^2).$$

Consider now the case where $\lambda(t) = \lambda_0 + a(R(t) - R_0)$ and $Q(t) = Q_0 + a(\gamma(t) - \gamma_0)$ where, of course, $R(t)$ and $\lambda(t)$ are two-state Markov processes such that $R(t) = R_d$ and $\gamma(t) = \gamma_d$ when $X(t) = d$, and $R(t) = R_p$ and $\gamma(t) = \gamma_p$ when $X(t) = p$. After simple but tedious calculations, we get

$$c_0 = \frac{Q_0}{\lambda_0} + a^2 \frac{p_d\tau_d(\gamma_d - \gamma_0 - Q_0 \dfrac{R_d - R_0}{\lambda_0})(R_0 - R_d)}{\lambda_0 + \tau_d p_p \lambda_0^2} + O(a^3) \qquad (80)$$

$$\mathrm{Var}(c(0)) = a^2 \frac{p_d\tau_d(\gamma_d - \gamma_0 - Q_0 \dfrac{R_0 - R_d}{\lambda_0})^2}{\lambda_0 + \tau_d p_p \lambda_0^2} + O(a^3) \qquad (81)$$

and

$$V^2_{c(0)} = a^2 \frac{p_d\tau_d\lambda_0(\dfrac{\gamma_d - \gamma_0}{Q_0} - \dfrac{R_d - R_0}{\lambda_0})^2}{1 + \tau_d p_p \lambda_0} + O(a^3).$$

To our knowledge the only exact calculations on models with dependence between the sink and source are due to Stein (1984). He considers a much more general situation very related to the (intensity) renewal model mentioned shortly in section 2. Thus (74) and (76) are special cases of his results, but due to the complexity of his model, we have chosen to present direct derivations. As an illustration of the generality of his results, we consider the case where $\lambda_d = 0$ and the length $\tilde{T}_d$ of a d-period may have an arbitrary distribution with variance σ_d^2. Then

$$c_0 = \frac{Q_0}{\lambda_0} + \frac{Q_d p_d(\sigma_d^2 + \tau_d^2)}{2\tau_d} \qquad (82)$$

and

$$\mathrm{Var}(c(0)) = Q_d^2 p_d \tau_d \{ \tau_d + \frac{1}{\lambda_0} + p_p\tau_d +$$

$$+ \frac{\sigma_d^2 - \tau_d^2}{2\tau_d^2}(\frac{1}{\lambda_0} - 2\tau_d p_d) - (\frac{\sigma_d^2 - \tau_d^2}{2\tau_d^2})^2 \tau_d p_d + (\frac{E(\tilde{T}_d^3)}{3\tau_d^2} - 2\tau_d)\}. \qquad (83)$$

It is seen that these formulae generalize (75) and (77). If $Q_d = Q_0$, i.e., if the source is deterministic, (82) follows from (14) by (7).

The models considered here do not include the case with independent sink and source unless one of them is deterministic. We shall now consider a point process version of the Markov model which also includes the case when the sink and the source are both random and independent.

Let $N_u(t)$ – here u stands for underlying – be a Poisson process, and let τ_u be the mean time between jumps in $N_u(t)$. To each jump we associate two random vectors (X,Y) and $(\tilde{\lambda},\tilde{Q})$ such that X and Y are zero-or-one variables and $\tilde{\lambda}$ and $\tilde{Q}$, as before, positive variables. Further we assume that (X,Y) and $(\tilde{\lambda},\tilde{Q})$ are independent of each other and of $N_u(t)$. The only allowed dependence is thus between X and Y and between $\tilde{\lambda}$ and $\tilde{Q}$. The sink and source processes are now defined by

$$\Lambda(t) = \sum_{k=1}^{N_u(t)} X_k\tilde{\lambda}_k \quad \text{and} \quad q(t) = \sum_{k=1}^{N_u(t)} Y_k\tilde{Q}_k$$

for $t \geq 0$ and in the obvious way for $t < 0$. The vectors (X_k,Y_k) and $(\tilde{\lambda}_k,\tilde{Q}_k)$, $k = 0,\pm1,\pm2,\ldots$, are independent of each other and of $N_u(t)$ and have the same distribution as (X,Y) and $(\tilde{\lambda},\tilde{Q})$.

Due to the properties of the Poisson process it follows that $\Lambda(t)$ is a Poisson model with $\tau_d = \tau_u/\Pr\{X = 1\}$ and that $q(t)$ is a Poisson model with $\tau_q = \tau_u/\Pr\{Y = 1\}$. Further $\Lambda(t)$ and $q(t)$ are independent if and only if $\Pr\{X = 1, Y = 1\} = 0$, i.e., if and only if they have no simultaneous jumps. As before $\tilde{\lambda}$ and $\tilde{Q}$ are the sizes of a typical jump in $\Lambda(t)$ and $q(t)$ respectively. Recall that $\tilde{Q}$ is the concentration emitted into the atmosphere. In the case of simultaneous jumps it might be more natural to consider $\tilde{Q}_e$, compare the mathematical remark in section 3, defined by

$$\tilde{Q} = \tilde{Q}_e(1 - e^{-X\tilde{\lambda}})/(X\tilde{\lambda}) = \tilde{Q}_e((1 - e^{-\tilde{\lambda}})/\tilde{\lambda})^X$$

as a model for the concentration emitted from the earth. Define, as before, $\phi(u) = E(e^{-u\tilde{\lambda}})$, and note that

$$E(e^{-uX\tilde{\lambda}}) = \Pr\{X = 0\} + \Pr\{X = 1\}\phi(u).$$

Consider now the concentration process $c(t)$. Since "nothing happens" between the jumps of N_u, it follows that $c(t)$ is constant between the jumps. If N_u has a jump at time t we have, compare again the mathematical remark in section 3,

$$c(t) = c(t-)e^{-X\tilde{\lambda}} + Y\tilde{Q}, \tag{84}$$

where (X,Y) and $(\tilde{\lambda},\tilde{Q})$ correspond to that jump and are independent of $c(t-)$. Since $c(t)$ is stationary it follows from (84) that

$$c_0 = c_0 \, E(e^{-X\tilde{\lambda}}) + E(Y\tilde{Q})$$

and thus

$$c_0 = \frac{E(Y\tilde{Q})}{1-E(e^{-X\tilde{\lambda}})} = \frac{\Pr\{Y=1\}E(\tilde{Q})}{\Pr\{X=1\}(1-\phi(1))} = \frac{(\tau_u/\tau_q)E(\tilde{Q})}{(\tau_u/\tau_d)(1-\phi(1))} = \frac{Q_0\tau_d}{1-\phi(1)}. \tag{85}$$

From (84) we get

$$c^2(t) = c^2(t-)e^{-2X\tilde{\lambda}} + Y\tilde{Q}^2 + 2c(t-)e^{-X\tilde{\lambda}}Y\tilde{Q}$$

since $Y^2 = Y$ and with similar arguments as above we get after some calculations

$$\mathrm{Var}(c(0)) =$$

$$= \frac{Q_0\tau_d^2}{1-\phi(1)} \left\{ \frac{2\Pr\{X=0|Y=1\}Q_0 + 2\Pr\{X=1|Y=1\}E(\tilde{Q}e^{-\tilde{\lambda}})/\tau_q}{1-\phi(2)} - \frac{Q_0}{1-\phi(1)} \right\} + $$

$$+ \frac{E(\tilde{Q}^2)}{\tau_q} \frac{\tau_d}{1-\phi(2)}. \tag{86}$$

At a first glance, (85) may seem surprising since it is unchanged whether Λ and q are independent or not. One explanation is that $\tilde{Q}$ is used instead of $\tilde{Q}_e$ and that $\tilde{Q} = \tilde{Q}_e$ if and only if Λ and q are independent

The concentration process is deterministic if $c(t) = \tilde{c}$. Then (84) implies that

$$\tilde{c} = \tilde{c}\, e^{-X\tilde{\lambda}} + Y\tilde{Q},$$

which holds if either $(X,Y) = (0,0)$ or if $(X,Y) = (1,1)$ and

$$\tilde{c}(1 - e^{-\tilde{\lambda}}) = \tilde{Q} = \tilde{Q}_e(1 - e^{-\tilde{\lambda}})/\tilde{\lambda}.$$

Thus $c(t)$ is deterministic if $X = Y$ and $\tilde{Q}_e = \tilde{c}\,\tilde{\lambda}$, which corresponds to the condition for intensity models.

We shall now indicate the flexibility of this model by considering some special cases. Before doing that we note that the distribution of $c(t)$ only depends on the sizes of the jumps of Λ and q and not on the positions of them. Thus (85) and (86) are true for any underlying point process N_u such that $\lim_{t\to-\infty} N_u(t) = -\infty$. Although we shall use this observation in "case 4," we shall in the other cases indicate that it is not as useful as might first be imagined.

Case 1 (Independent Poisson sink and source)

Choose τ_u such that $1/\tau_u = 1/\tau_d + 1/\tau_q$ and put

$$\Pr\{(X,Y) = (i,j)\} = \begin{cases} \tau_u/\tau_d \text{ if } (i,j) = (1,0) \\ \tau_u/\tau_q \text{ if } (i,j) = (0,1) \\ 0 \text{ otherwise.} \end{cases}$$

This implies that $\Pr\{X=0|Y=1\} = 1$ and thus (86) reduces to (34).

Assume now that N_u is not a Poisson process. If there, as above, are no simultaneous jumps, then it follows from Matthes et al. (1978, p.388) that Λ and q are dependent. If $\Pr\{(X,Y) = (1,1)\} > 0$ then Λ and q have simultaneous jumps and thus they are dependent if $\Pr\{N_u(t) = N_u(t-)\} = 0$ for all values of t, which is the case if N_u is stationary. Thus, Λ and q are stationary and independent if and only if they are Poisson models.

Case 2 (Poisson sink and deterministic source)

Consider the case when $Y = 1$ and $\tilde{Q} = Q_0\tau_u$. Then $q(t) = Q_0\tau_u N_u(t)$ and thus $E(q(t)) = Q_0\tau_u t/\tau_u = Q_0 t$ and $\mathrm{Var}(q(t)) = Q_0^2\tau_u^2 t/\tau_u = \tau_u Q_0^2 t$.

Thus it seems reasonable to believe that we are close to a deterministic source if τ_u is small. More precisely, we consider a sequence $\tau_u^{(n)} \downarrow 0$ and the corresponding sequence $N_u^{(n)}$ of Poisson processes. Then

$$\tau_u^{(n)} N_u^{(n)}(t) \overset{p}{\to} t \quad \text{as } n \to \infty, \tag{87}$$

where $\overset{p}{\to}$ means "convergence in probability," and thus

$$q^{(n)}(t) = Q_0 \tau_u^{(n)} N_u^{(n)}(t) \overset{p}{\to} Q_0 t.$$

Put $Y = 1$, $\tilde{Q} = Q_0\tau_u$ and $\Pr\{X = 1\} = \tau_u/\tau_d$. Then (86) reduces to

$\mathrm{Var}(c(0))=$

$$= \frac{Q_0\tau_d^2}{1-\phi(1)} \left\{ \frac{2(1 - \tau_u/\tau_d)Q_0 + 2(\tau_u/\tau_d)Q_0\tau_u\phi(1)/\tau_q}{1 - \phi(2)} - \frac{Q_0}{1-\phi(1)} \right\} + \frac{Q_0^2\tau_u\tau_d}{1-\phi(2)}$$

which reduces to (31) when $\tau_u \to 0$. It shall be emphasized that this is not an alternative proof of (31) since convergence in probability does not automatically imply convergence of moments.

Assume now that $N_u^{(n)}$ are point processes, but not necessarily Poisson processes. Assume that there exists a sequence $\tau_u^{(n)}$ such that (87) holds. Let $\Lambda^{(n)}$ be the corresponding sink process; i.e.,

$$\Lambda^{(n)}(t) = \sum_{k=1}^{N_u^{(n)}(t)} \tilde{X}_k^{(n)} \tilde{\lambda}_k^{(n)}$$

for $t > 0$, and let $N^{(n)}$ be the point process corresponding to $\Lambda^{(n)}$; i.e.,

$$N^{(n)}(t) = \sum_{k=1}^{N_u^{(n)}(t)} \tilde{X}_k^{(n)}.$$

Put $p^{(n)} = \Pr\{\tilde{X}^{(n)} = 1\}$. Assume there exists a sequence $p^{(n)} \downarrow 0$ and a point process N, not identically equal to zero, such that

$$N^{(n)} \overset{d}{\to} N \text{ as } n \to \infty. \tag{88}$$

where $\overset{d}{\to}$ means "convergence in distribution". (The distribution of a point process is a probability measure on the set of all realizations of point processes. Endowed with the vague topology that set is a Polish space.) We shall now show that if (87) and (88) hold, then N must be a Poisson process.

It follows from Kallenberg (1975, p 57) that (88) holds if and only if there exists a random measure (a process with non-negative increments) η such that

$$p^{(n)}N_u^{(n)} \overset{d}{\to} \eta \text{ as } n \to \infty. \tag{89}$$

Further N is a Cox process directed by η, i.e. $N(t) = \tilde{N}(\eta(t))$ where $\tilde{N}$ is a Poisson process with $E(\tilde{N}(t)) = t$ which is independent of η. Thus it remains to show that $\eta(t) = \rho t$ for some $\rho \in (0,\infty)$.

Let T_η consist of those t for which $Pr\{\eta(t) - \eta(t-) = 0\} = 1$. For each $t \in T_\eta$ it follows from (89) that

$$p^{(n)}N_u^{(n)}(t) \overset{d}{\to} \eta(t) \text{ as } n \to \infty.$$

Recall (87) and put $\rho^{(n)} = p^{(n)}/\tau_u^{(n)}$. There exist a subsequence $\{n'\}$ such that $\rho^{(n')} \to \rho \in [0,\infty]$. Then

$$\eta(t) \overset{d}{\leftarrow} p^{(n')}N_u^{(n')}(t) = \rho^{(n')}\tau_u^{(n')}N_u^{(n')}(t) \overset{p}{\to} \rho t$$

and thus $\eta(t) = \rho t$. Since $\eta(t) < \infty$ we have $\rho < \infty$. Since T_η is dense in $(-\infty,\infty)$ we have $\eta(t) = \rho t$ for all t. If $\rho = 0$ then $Pr\{N(t) \equiv 0\} = 1$, which contradicts (86), and thus $\rho \in (0,\infty)$.

Case 3 (Point process correspondence to the Markov model)

Consider the intensity model, discussed in the beginning of this appendix, where the underlying process X(t) is a two-state Markov process taking the values d and p. Let $\tilde{T}_d$ and $\tilde{T}_p$ be the length of a typical d-period and a typical p-period respectively. Thus $\tilde{T}_d$ ($\tilde{T}_p$) is exponentially distributed with mean τ_d (τ_p).

A natural point process correspondence is when $\tau_u = \tau_d$, $X = Y = 1$, $\tilde{\lambda} = \lambda_p \tilde{T}_p$ and $\tilde{Q}_e = Q_p \tilde{T}_p$ which corresponds to an intensity model with $\lambda_d = Q_d = 0$. This choice does, however, yield a deterministic concentration process and is thus of limited interest.

Consider again the intensity model but this time we assume that $Q_d > 0$, in order to avoid a deterministic concentration process, and that $\lambda_d = 0$, since then the formulae are very much simplified. In the point process model we put, compare case 2, $X = Y = 1$ and

$$(\tilde{\lambda},\tilde{Q}_e) = \begin{cases} (0, q_d \tau_u) & \text{with probability } 1 - \tau_u/\tau_d \\ (\mu_\lambda Z, \mu_Q Z) & \text{with probability } \tau_u/\tau_d \end{cases}$$

where Z is an exponentially distributed random variable with mean 1 and where q_d, μ_λ and μ_Q are constants. Since $\tilde{T}_p = \tau_p Z$ the intuitively natural choice is $q_d = Q_d$, $\mu_\lambda = \lambda_p \tau_p$ and $\mu_Q = Q_p \tau_p$. One drawback with this choice is that the means of the sink and the source are different in the intensity model and in the point process model. Let λ_0 and Q_0 be the means in the intensity model. For the sink process we have

$$E(\Lambda(1)) = (1/\tau_u)(\tau_u/\tau_d)\mu_\lambda = \mu_\lambda/\tau_d$$

and thus we choose $\mu_\lambda = \lambda_0 \tau_d$, since then $E(\Lambda(1)) = \lambda_0$. The same argument for the source process leads to the relation

$$(\tau_d - \tau_u) q_d + \mu_Q = Q_0 \tau_d,$$

which simplifies to

$$q_d + \mu_Q/\tau_d = Q_0 \tag{90}$$

when $\tau_u \to 0$ as we shall let it do. Thus we put $\mu_\lambda = \lambda_0 \tau_d$ and assume that (90) holds.

When $\tau_u \to 0$ it follows after some calculations that (85) and (86) reduce to

$$c_0 = \frac{Q_0}{\lambda_0} + q_d\tau_d \tag{91}$$

and

$$\mathrm{Var}(c(0)) = q_d^2\tau_d(\tau_d + \lambda_0^{-1})$$

which shall be compared with (75) and (77).

Certainly it is not enough to require that (90) holds, since we also want the point process model to be "close" to the intensity model. One natural choice is

$$q_d = Q_d \quad \text{and} \quad \mu_Q = \tau_d(Q_0 - Q_d) = p_d\tau_p(Q_p - Q_d) \tag{92}$$

since then the two models coincide during d-periods. This choice does, however, only work if $Q_p \geq Q_d$. If $Q_p = Q_d$ the source is deterministic in the intensity model and (92) also yields a deterministic source in the point process model. We note that the point process model tends to the S. R. Markov model mentioned in section 4 as $\tau_u \to 0$. In our opinion "the most natural" choice is

$$q_d = Q_d \quad \text{and} \quad \mu_Q = p_d\tau_p Q_p \tag{93}$$

for at least three reasons. Firstly, the source is equally modified in d- and p-periods in the sense that $q_d\tau_d/\mu_Q = Q_d\tau_d/(Q_p\tau_p)$. Secondly, $q_d = p_d Q_d$ is the q_d-value closest to Q_d for which $\mu_Q \geq 0$ for all values of Q_p. Thirdly, (91) coincides with (75).

<u>Case 4</u> (Point process correspondence to Stein´s model)

We have already referred to a special case of the model due to Stein (1984). In the general formulation his model is characterized by two random vectors $(\tilde{T}_d,\tilde{\lambda}_d,\tilde{Q}_d)$ and $(\tilde{T}_p,\tilde{\lambda}_p,\tilde{Q}_p)$ where $\tilde{T}_d$ is the length of a typical d-period, $\tilde{\lambda}_d$ the sink intensity and $\tilde{Q}_d$ the source strength of that period. The vector $(\tilde{T}_p,\tilde{\lambda}_p,\tilde{Q}_p)$ has the same interpretation for a typical p-period. Stein´s model is a renewal model in the sense that the characteristics of different periods are independent of each other. The Markov model considered in this appendix is thus the particular case where $\tilde{\lambda}_p = \lambda_p$, $\tilde{Q}_p = Q_p$, $\tilde{\lambda}_d = \lambda_d$, $\tilde{Q}_d = Q_d$ and where $\tilde{T}_d$ and $\tilde{T}_p$ are exponentially distributed.

Assume now that $\tilde{T}_d$ is not exponentially distributed, but that $\tilde{Q}_d = \tilde{\lambda}_d = 0$. This means that "nothing happens" in the d-periods. In the point process model we put $X = Y = 1$ and $(\tilde{\lambda},\tilde{Q}_e) = (\tilde{T}_p\tilde{\lambda}_p,\tilde{T}_p\tilde{Q}_p)$. Thus c_0 and $\mathrm{Var}(c(0))$ follow from (85) and (86). Certainly this is not a very interesting special case of Stein's model, and it is merely used as an illustration of a case, where the observation that N_u does not need to be a Poisson process, is applicable.

Stein considers the special case where $\tilde{T}_d$ and $\tilde{T}_p$ are exponentially distributed, $\tilde{Q}_d = \tilde{Q}_p = Q_0$, $\tilde{T}_d$ and $\tilde{\lambda}_d$ are independent and $\tilde{T}_p$ and $\tilde{\lambda}_p$ are independent. If we further assume that $\tilde{\lambda}_d = 0$ this case corresponds to the Poisson model with $\tilde{\lambda} = \tilde{T}_p\tilde{\lambda}_p$ and we have

$$c_0 = Q_0\tau_d/(1 - \phi(1)).$$

Recall that $c_0 \geq Q_0/\lambda_0$, see (10) or appendix A1, which means that c_0 is minimized if the sink is deterministic. The following variation of this inequality is essentially due to Stein (1984). Let $\tilde{T}_p$ have an arbitrary distribution. Define the Laplace-transform

$$\hat{f}_p(u) = E(e^{-u\tilde{T}_p})$$

and thus we have

$$\phi(1) = E(\hat{f}_p(\tilde{\lambda}_p))$$

Since a Laplace-transform is convex it follows from Jensen's inequality, see appendix A1, that

$$\phi(1) \geq \hat{f}_p(\lambda_p)$$

where $\lambda_p = E(\tilde{\lambda}_p)$. Thus, in the Poisson model, c_0 is minimized if $\tilde{\lambda}_p = \lambda_p$. (By symmetry it follows that if $\tilde{\lambda}_p$ is kept then c_0 is minimized if $\tilde{T}_p = \tau_p$). If $\tilde{T}_d$ has an arbitrary distribution we have the S. R. renewal model, and it follows from (17) that the conclusion still holds.

Let us now return to the two-state Markov intensity model and consider the distribution of the random variable $c(0)$. Define the Laplace-transforms

$$\phi_{c(0)}(u) = E(e^{-uc(0)}),$$

$$\phi_d(u) = E(e^{-uc(0)} | X(0) = d) \text{ and } \phi_p(u) = E(e^{-uc(0)} | X(0) = p)$$

and note that

$$\phi_{c(0)}(u) = p_d\phi_d(u) + p_p\phi_p(u).$$

Let the random variable c_d (c_p) be the concentration at a time when $X(t)$ has a jump $d \to p$ ($p \to d$). Due to the "lack of memory" of the exponential distribution it follows that

$$E(e^{-uc_d}) = \phi_d(u) \quad \text{and} \quad E(e^{-uc_p}) = \phi_p(u)$$

Consider now a jump $p \to d$ of $X(t)$ and let c_p be the concentration at that jump. Let c_d be the concentration at the subsequent jump $d \to p$ of $X(t)$. Let $\tilde{T}_d$ be the length of the d-period between those two jumps. Thus we have

$$c_d = c_p e^{-\lambda_d \tilde{T}_d} + Q_d \int_0^{\tilde{T}_d} e^{-\lambda_d(\tilde{T}_d - s)} ds = (c_p - \frac{Q_d}{\lambda_d})e^{-\lambda_d \tilde{T}_d} + \frac{Q_d}{\lambda_d}$$

or

$$c_d - \frac{Q_d}{\lambda_d} = (c_p - \frac{Q_d}{\lambda_d})e^{-\lambda_d \tilde{T}_d} .$$

Since $\tilde{T}_d$ is exponentially distributed with mean τ_d and since $\tilde{T}_d$ and c_p are independent we get

$$\phi_d(u)e^{-uQ_d/\lambda_d} = \int_0^\infty \phi_p(ue^{-\lambda_d x}) \exp(u \frac{Q_d}{\lambda_d} e^{-\lambda_d x}) \frac{1}{\tau_d} e^{-x/\tau_d} dx.$$

Simple calculations and differentiation, cf. the mathematical remark in section 8, yield

$$u\phi_d'(u)\tau_d\lambda_d + u\phi_d(u)\tau_d Q_d + \phi_d(u) = \phi_p(u). \tag{94}$$

By completely analogous arguments it follows that

$$u\phi_p'(u)\tau_p\lambda_p + u\phi_p(u)\tau_p Q_p + \phi_p(u) = \phi_d(u). \tag{95}$$

We have not managed to solve this system of differential equations explicitly. By expansion of the Laplace-transforms we get, for $k \geq 1$,

$$E(c_d^k)(1 + k\tau_d\lambda_d) - E(c_p^k) = E(c_d^{k-1})k\tau_d Q_d \tag{96}$$

$$-E(c_d^k) + E(c_p^k)(1 + k\tau_p\lambda_p) = E(c_p^{k-1})k\tau_p Q_p. \tag{97}$$

Using this we can compute all moments of c(0) and we have an alternative to the method used in the derivation of (73) and (76). The formula for all moments of c(0) stated – without proof – by Baker et al. (1984, p 970) for a deterministic source follows easily from (96) and (97).

Now we consider the case $\lambda_d = 0$. Then (94) reduces to

$$\phi_d(u) = \phi_p(u)(1 + u\tau_d Q_d)^{-1} \tag{98}$$

and thus it follows from (95) that

$$u\phi_p'(u)\tau_p\lambda_p = \phi_p(u)\{(1 + u\tau_d Q_d)^{-1} - u\tau_p Q_p - 1\}$$

or

$$\phi_p'(u) = \phi_p(u)\{ \frac{\tau_d Q_d}{\tau_p\lambda_p(1 + u\tau_d Q_d)} - \frac{Q_p}{\lambda_p} \}.$$

Thus, cf. the mathematical remark in section 8,

$$\phi_p(u) = e^{-uQ_p/\lambda_p}(1 + u\tau_d Q_d)^{-1/(\tau_p\lambda_p)}. \tag{99}$$

From (98) and (99) we get

$$\phi_{c(0)}(u) = e^{-uQ_p/\lambda_p}(1 + u\tau_d Q_d)^{-1/(\tau_p\lambda_p)}(p_p + \frac{p_d}{1 + u\tau_d Q_d}). \tag{100}$$

For reference reasons we say that non-negative random variable Z is
Exp-zero(β,p)-distributed if it has distribution function

$$F_Z(z) = \begin{cases} 0 & \text{if } z < 0 \\ 1 - (1-p)e^{-z/\beta} & \text{if } z \geq 0. \end{cases}$$

This means that Z is zero with probability p and exponentially
distributed with probability 1-p.

Thus

$$c(0) \overset{d}{=} \frac{Q_p}{\lambda_p} + Y + Z \tag{101}$$

where Y is $\Gamma((\tau_p\lambda_p)^{-1}, \tau_d Q_d)$-distributed, cf. section 8, and Z is
Exp-zero($\tau_d Q_d$, p_p)-distributed. Further Y and Z are independent.

Consider now "long-lived" particles. Then

$$c(0) \overset{d}{=} \frac{\gamma_p}{R_p} + Y + Z \tag{102}$$

where Y and Z are independent, Y is $\Gamma((a\tau_p R_p)^{-1}, a\tau_d\gamma_d)$- and
Z is Exp-zero($a\tau_d\gamma_d$, p_p)-distributed. Define, like in section 8,
$\bar{c}_a(0)$ by

$$\bar{c}_a(0) = (c(0) - \gamma_0/R_0)/\sqrt{a}.$$

Thus we have

$$\bar{c}_a(0) \overset{d}{=} (\frac{\gamma_p}{R_p} - \frac{\gamma_0}{R_0} + Y + Z)/\sqrt{a} = (Y - \frac{\tau_d\gamma_d}{\tau_p R_p})/\sqrt{a} + Z/\sqrt{a}.$$

Since $\Pr\{Z/\sqrt{a} > \varepsilon\} = p_d \exp(-\varepsilon/(\tau_d\gamma_d\sqrt{a})) \to 0$ as $a \to 0$ it follows
from the gamma approximation, cf. section 8, that

$$\bar{c}_a(0) \overset{d}{\to} \frac{\tau_d\gamma_d}{\sqrt{\tau_p R_p}} \cdot W \tag{103}$$

where W, as before, is a normally distributed random variable, with
E(W) = 0 and Var(W) = 1. For a deterministic source (103) follows
from the approximation discussed in section 8.

Consider now the point process model. We have not managed to derive the distribution of c(0) explicitly except when the sink and the source are independent and where $\tilde{\lambda}$ and $\tilde{Q}$ are exponentially distributed. Otherwise expressed we shall consider the case when the sink and the source are described by independent S. R. Markov models. We shall heavily rely on the intensity model, and therefore we consider the case here instead of in section 8 where it might logically belong.

Choose τ_u as in case 1 and recall that a jump in N_u corresponds to a jump in Λ with probability $\tau_u/\tau_d = \tau_q/(\tau_d + \tau_q)$ and to a jump in q with probability $\tau_u/\tau_q = \tau_d/(\tau_d + \tau_q)$.

The idea is to construct an intensity model such that c(0) is the same as in the point process case and such that (101) is applicable. In order to keep the models apart, we add an index "i" to all quantities related to the intensity model. Recall that c(0) only depends on the sizes of the jumps of $N_u(t)$ for $t < 0$ and not on their positions. Thus c(0) is completely determined by the vectors $(X_k\tilde{\lambda}_k, Y_k\tilde{Q}_k)$ for $k = 0,-1,-2,\dots$.

Consider a vector process $X^i(t) = (\lambda^i(t), Q^i(t))$, $t \le 0$ which alternates between the states $p^i = (\lambda^i_p, 0)$ and $d^i = (0, Q^i_d)$. Put

$$S^i_k = (X_k\tilde{\lambda}_k/\lambda^i_p) + (Y_k\tilde{Q}_k/Q^i_d)$$

and

$$(\lambda^i(t), Q^i(t)) = (X_k\lambda^i_p, Y_kQ^i_d) \quad \text{if} \quad -\sum_{j=k}^{0} S^i_j < t < -\sum_{j=k+1}^{0} S^i_j$$

for $k = 0,-1,-2,\dots$. Recall the convention that $\sum_{1}^{0} \dots = 0$. This means that the "effect" of a jump in N_u is "spread out" over an interval. The parameters λ^i_p and Q^i_d will be choosen later. It is not difficult to realize that $c^i(0) = c(0)$. Thus it remains to show that (101) is applicable; i.e., that $X^i(t)$ is a stationary Markov process. Assume that $X^i(0) = p^i$ which holds if $(X_0, Y_0) = (1,0)$. The length $\tilde{T}^i_p$ of that p^i-period is equal to $\tilde{\lambda}_0/\lambda^i_p$ if $(X_{-1}, Y_{-1}) = (0,1)$ and equal to $(\tilde{\lambda}_0 + \tilde{\lambda}_{-1} + \dots + \tilde{\lambda}_k)/\lambda^i_p$ if $(X_{-1}, Y_{-1}) = \dots = (X_k, Y_k) =$

$= (1,0)$ and $(X_{k-1},Y_{k-1}) = (1,0)$. Since the $\tilde{\lambda}_k$:s are independent of each other and of the (X_k,Y_k):s and exponentially distributed with mean $\tau_d\lambda_0$ we have

$$E(e^{-u\tilde{T}^i_p}) = \sum_{k=1}^{\infty} (1 + u\tau_d\lambda_0/\lambda^i_p)^k \cdot (\tau_q/(\tau_d + \tau_q))^{k-1} \cdot (\tau_d/(\tau_d + \tau_q)) =$$

$$= (1 + u(\tau_d + \tau_q)\lambda_0/\lambda^i_p)^{-1}.$$

Thus $\tilde{T}^i_p$ is exponentially distributed with mean $\tau^i_p = (\tau_d+\tau_q)\lambda_0/\lambda^i_p$. By analogous arguments it follows that the length of the preceding d^i-period is exponentially distributed with mean $\tau^i_d = (\tau_d+\tau_q)Q_0/Q^i_d$. The two periods are further independent of each other since they are formed by independent variables. Proceeding in this way we realize that $X^i(t)$ is a Markov process. Since $\Pr\{X^i(0) = p^i\} = \Pr\{(X_0,Y_0) = (1,0)\} = \tau_q/(\tau_d + \tau_q)$ it follows that $X^i(t)$ is stationary if $\tau^i_d/(\tau^i_d+\tau^i_p) = \tau_q/(\tau_d+\tau_q)$ which holds for $\lambda^i_p = \lambda_0/\tau_d$ and $Q^i_d = Q_0/\tau_q$. Thus (101) is applicable and it follows that

$$c(0) \overset{d}{=} Y + X \tag{104}$$

where Y and Z are independent, Y is $\Gamma(((\tau_d+\tau_q)\lambda_0)^{-1},(\tau_d+\tau_q)Q_0)$-distributed and Z is Exp-zero$((\tau_d+\tau_q)Q_0,\tau_q/(\tau_d+\tau_q))$-distributed. For "long-lived" particles it follows from (103) that

$$\bar{c}_a(0) \overset{d}{\to} \gamma_0\sqrt{(\tau_d + \tau_q)/R_0} \cdot W \text{ as } a \to 0.$$

A5 PROOF OF FORMULA (38)

Let W_f be the distance from time x to the next "shower" and W_b be the distance from x to the previous one. The definitions are illustrated in Figure 8, where the crosses indicate the time of the "showers." Note that $x, y \leq 0$ while $W_f, W_b \geq 0$.

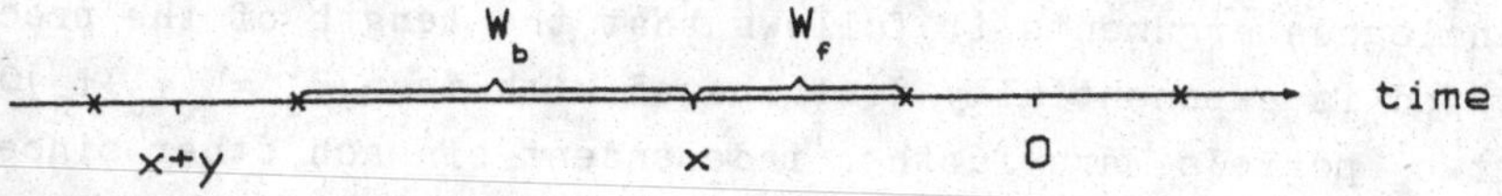

FIGURE 8: Illustration of notation.

Let $G_p(t)$ be the survivor function for a particle which enters the atmosphere immediately before a "shower." Put $G_p(t) = 1$ for $t < 0$ but note that $G_p(0+) = \lim_{t \downarrow 0} G_p(t) = \phi(1)$. Let $G_p^{(2)}(t)$ be the corresponding survivor function when $\Lambda(t)$ is replaced by $2\Lambda(t)$. From the mathematical remark in section 4 it follows that

$$E_p(T^{(2)}) = \int_0^\infty G_p^{(2)}(t)\,dt = \tau_d \phi(2)/(1 - \phi(2)).$$

Let us now introduce some notation. Let $F_d(x)$ and $f_d(x)$ be the distribution function and the density respectively of the time between two "showers" and put

$$\hat{f}_d(u) = \int_0^\infty e^{-ux} f_d(x)dx.$$

This implies, compare section 2, that

$$E(e^{-uW_b}) = \frac{1}{\tau_d} \int_0^\infty e^{-ux}(1 - F_d(x))dx = (1 - \hat{f}_d(u))/(u\tau_d)$$

and that, cf. Feller (1971, p. 386),

$$E(W_f e^{-uW_b}) = \frac{1}{\tau_d} \int_0^\infty \int_0^\infty y e^{-ux} f_d(x+y)dydx =$$

$$= \frac{1}{\tau_d} \int_0^\infty e^{-ux} \int_x^\infty (1 - F_d(y))dydx =$$

$$= (1 - E(e^{-uW_b}))/u = (\hat{f}_d(u) - 1 + u\tau_d)/(u^2\tau_d).$$

Finally we put

$$\hat{G}_p(u) = \int_0^\infty e^{-ut} G_p(t)dt.$$

From all this we get

$$B = 2\sigma_Q^2 \int_{-\infty}^0 \int_{-\infty}^0 E(e^{2\Lambda(x)} e^{\Lambda(x+y)-\Lambda(x)}) e^{A_Q y} \, dydx =$$

$$= 2\sigma_Q^2 \int_{-\infty}^0 \int_{-\infty}^0 E(G_p^{(2)}(-x - W_f)G_p(-y - W_b))e^{\Lambda_Q y} dydx =$$

$$= 2\sigma_Q^2 E(\{W_f + E_p(T^{(2)})\}\{ \int_{-W_b}^0 e^{A_Q y} dy + \int_{-\infty}^{-W_b} G_p(-y - W_b)e^{A_Q y} dy\}) =$$

$$= 2\sigma_Q^2 \; E(\{W_f + E_p(T^{(2)})\}\{(1 - e^{-A_Q W_b})/A_Q + e^{-A_Q W_b} \hat{G}_p(A_Q)\}),$$

and thus we need an expression for $\hat{G}_p(A_Q)$. By an ordinary "renewal argument," it follows that $G_p(t)$ satisfies the renewal equation

$$G_p(t) = \phi(1)(1 - F_d(t)) + \phi(1) \int_0^t G_p(t-s)f_d(s) \, ds$$

and thus

$$\hat{G}_p(u) = \phi(1)(1 - \hat{f}_d(u))/u + \phi(1)\hat{G}_p(u)\hat{f}_d(u)$$

or

$$\hat{G}_p(A_Q) = \frac{\phi(1)(1 - \hat{f}_d(A_Q))}{A_Q(1 - \phi(1)\hat{f}_d(A_Q))} .$$

Thus we have

$$B = \frac{2\sigma_Q^2}{A_Q} \{E(T^{(2)}) + [E(W_f e^{-A_Q W_b}) + E_p(T^{(2)})E(e^{-A_Q W_b})](A_Q\hat{G}_p(A_Q) - 1)\} =$$

$$= \frac{2\sigma_Q^2}{A_Q} \{ \frac{\sigma_d^2 - \tau_d^2}{2\tau_d} + \frac{\tau_d}{1-\phi(2)} + \frac{1 - \phi(1)}{A_Q(1-\phi(1)\hat{f}_d(A_Q))}$$

$$(\frac{1-\hat{f}_d(A_Q)}{A_Q\tau_d} - \frac{1-\phi(2)\hat{f}_d(A_Q)}{1 - \phi(2)})\}$$

which was to be proved.

REFERENCES

Alexander, K. (1981). Determination of rainfall duration statistics for rainout models from daily records. Water Resour. Res. 17, 521-528.

Baker, M. B., Harrison, H., Vinelli, J. and Ericsson, K.B. (1979). Simple stochastic models for the sources and sinks of two aerosol types. Tellus 31, 39-51.

Baker, M. B., Eylander, M. and Harrison, H. (1984). The statistics of chemical trace concentrations in the steady state. Atmos. Environ. 18, 969 - 975.

Billingsley, P. (1968). Convergence of probability measures. Wiley, New York.

Blake, I. F. and Lindsey, W. C. (1973). Level-crossing problems for random processes. IEEE Trans. Inform. Theory. 19, 295 - 315.

Cramér, H. (1945). Mathematical methods of statistics. Almqvist and Wiksell, Stockholm and Princeton University Press, Princeton.

Daley, D.J. and Vere-Jones, D. (1972). A summary of the theory of point processes. Stochastic point processes: Statistical analysis, theory and applications. Ed. by Lewis, P.A.W., 299-383. Wiley - Interscience, New York.

Elliott, R.J. (1982) Stochastic calculs and applications. Springer-Verlag, New York.

Feller, W. (1971). An introduction to probability theory and its applications. Vol. II. 2nd. ed. John Wiley and Sons, New York

Gibbs, A. G. and Slinn, W. G. N. (1973). Fluctuations in trace gas concentrations in the troposphere. J. Geophys. Res. 78, 574-576.

Grandell, J. and Rodhe, H. (1978). A mathematical model for the residence time of aerosol particles removed by precipitation scavenging. Trans. 8th Prague Conf. A, 247-261.

Grandell, J. (1980). Approximate waiting times in thinned point processes. Liet. matem. rink. XX, No 4, 29-47.

Grandell, J. (1982). Mathematical models for the variation of air-pollutant concentrations. Adv. Appl. Prob. 14, 240-256.

Grandell, J. (1983a). Estimation of precipitation characteristics fromm time-integrated data. Trans 9th Prague Conf, 263-268.

Grandell, J. (1983b). Some remarks on the age distribution of air pollutants. Recent Trends in Mathematics. TEUBNER-TEXTE zur Mathematik 50. Teubner Verlagsgesellschaft, Leipzig.

Grenander, U. and Rosenblatt, M. (1956). Statistical analysis of stationary time series. Almqvist & Wiksell, Stockholm, and John Wiley and Sons, New York.

Hamrud, M., Rodhe, H. and Grandell, J. (1981). A numerical comparision between Lagrangian and Eulerian rainfall statistics. Tellus 33, 235-241.

Hamrud, M. (1983). Residence time and spatial variability for gases in the atmosphere. Tellus 35B, 295-303.

Junge, C.E. (1974). Residence time and variability of tropospheric trace gases. Tellus 26, 477-488.

Kallenberg, O. (1975). Random measures. Akademie-Verlag, Berlin and Academic Press, London.

Karr, A.F.(1984). Estimation and reconstruction for Zero-One Markov processes. Stochastic Process. Appl. 16, 219-256.

Kelly, F.P. (1979). Reversibility and stochastic networks. John Wiley and Sons, New York.

Leadbetter, M. R., Lindgren, G. and Rootzén, H. (1983). Extremes and related properties of random sequences and processes. Springer - Verlag, New York Heidelberg Berlin.

Lindvall, T. (1973). Weak convergence of probability measures and random functions in the function space D [0,∞). J. Appl. Prob. 10, 109 - 121.

Lozowsky, E. (1983). The spatial inhomogenity of aerosols within an air parcel and some implications for the modelling of particle scavenging, by convective clouds. Precipitation scavenging, dry deposition and resuspension. Ed. by Pruppacher, H.R. et al. Elsevier, New York.

Matthes, K., Kerstan, J. and Mecke, J. (1978). Infinitely divisible point processes. John Wiley and Sons, New York.

Rodhe, H. and Grandell, J. (1972). On the removal time of aerosol particles from the atmosphere by precipitation scavenging. Tellus 24, 442-454.

Rodhe, H. and Grandell, J. (1981). Estimates of characteristic times for precipitation scavenging. J. Atmos. Sc. 38, 370-386.

Slinn, W.G.N. (1982). Estimates for the long-range transport of air pollution. Water, Air and Soil Pollution 18, 45-64.

Stein, M. (1984). System parameters governed by jump processes: A model for removal of air pollutants. Adv. Appl. Prob. 16, 603-617.

SUBJECT INDEX

INDEX OF REFERENCES

INDEX OF NOTATION

General

X is a random variable

μ_X or E(X) mean value 4

σ_X standard deviation 4

σ_X^2 or Var(X) variance 4

V_X $= \sigma_X/\mu_X$ coefficient of variation 4

V_X^2 relative variance 4, 14

$F_X(x)$ $= \Pr\{X \leq x\}$ distribution function 4

$f_X(x)$ density function

$\hat{f}_X(u)$ or $\phi_X(u)$ Laplace-transform 19, 25

$G_X(x)$ $= \Pr\{X > x\}$ survivor function 4

X(t) is a stationary stochastic processs

$r_X(\tau)$ covariance function 4

$f_X(\omega)$ spectral density function 5

The sink

$\lambda(t)$ sink intensity 10

$\Lambda(t)$ sink process 12

λ_0 $= E(\lambda(t)) = E(\Lambda(1))$ 17

a $: \lambda(t) = aR(t)$ 10

R(t) precipitation intensity 10, 12

h(t) precipitation process 12

R_0 $= E(R(t)) = E(h(1))$ 20

τ_d 18, λ_d 18, $\tilde{T}_d$ 42 characteristics related to d-periods

τ_p 18, λ_p 18, $\tilde{T}_p$ 42 characteristics related to p-periods

$\tilde{\lambda}_k$ sink at "shower" 14, 19

$\tilde{M}_k$ precipitation at "shower" 42

Γ_h $= \lim_{t\to\infty} t^{-1}\mathrm{Var}(h(t))$ 20

The source

$Q(t)$ source strength 10, 12

$q(t)$ source process 12

Q_0 $= E(Q(t)) = E(q(1))$ 17

$\gamma(t)$ 38

$g(t)$ 28

γ_0 $= E(\gamma(t)) = E(g(1))$ 26, 28

τ_q, $\tilde{Q}_k$, $\tilde{T}_q$, N_q characteristics in the point process case 24

Γ_g $= \lim_{t\to\infty} t^{-1}\mathrm{Var}(g(t))$ 28

Pollutants

T or T_a residence time 14, 17, 20

$G(t)$ or $G_a(t)$ survivor function 14, 20

$c(t)$ concentration process 10, 13

c_0 $= E(c(t))$ 17

9 780387 961972